PAINLESS POLICE
REPORT WRITING

PAINLESS POLICE REPORT WRITING

An English Guide for Criminal Justice Professionals

Barbara Frazee
Joseph N. Davis

PRENTICE HALL
Englewood Cliffs, New Jersey 07632

Library of Congress Cataloging-in-Publication Data

Frazee, Barbara.
   Painless police report writing : an English guide for criminal
justice professionals / Barbara Frazee, Joseph N. Davis.
     p.   cm.
   Includes student workbook, answer key, and index.
   ISBN 0-13-647629-5
   1. Police reports.   2. English language—Rhetoric.   3. Police
reports.—United States.   I. Davis, Joseph N.   II. Title.
HV7936.R53F73   1993
808'.066364—dc20
                                        92-33194
                                        CIP

Acquisitions editor: Robin Baliszewski
Editorial assistant: Rose Mary Florio
Editorial/production supervision and
   interior design: Marcia Krefetz/Linda B. Pawelchak
Copy editor: William O. Thomas
Cover design: Mike Fender
Prepress buyer: Ilene Levy
Manufacturing buyer: Ed O' Dougherty

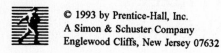

© 1993 by Prentice-Hall, Inc.
A Simon & Schuster Company
Englewood Cliffs, New Jersey 07632

Printed in the United States of America
10 9 8 7 6

ISBN 0-13-647629-5

Prentice-Hall International (UK) Limited, *London*
Prentice-Hall of Australia Pty. Limited, *Sydney*
Prentice-Hall Canada Inc., *Toronto*
Prentice-Hall Hispanoamericana, S.A., *Mexico*
Prentice-Hall of India Private Limited, *New Delhi*
Prentice-Hall of Japan, Inc., *Tokyo*
Simon & Schuster Asia Pte. Ltd., *Sinapore*
Editora Prentice-Hall do Brasil, Ltda., *Rio de Janeiro*

CONTENTS

Chapter Six

Chapter Seven

PREFACE

Legal charges may be filed or dismissed; court cases may be won or lost—all based on police officers' reports. Most law enforcement officers do not go into law enforcement because they love to write reports. When officers do encounter writing problems, often they do not know where to turn to find the solutions to these problems. Police academy report writing classes may be taught by law enforcement professionals without the English grammar to explain concepts and help struggling recruits. In desperation, frustrated officers turn to lower division and community college English composition classes.

Unfortunately, traditional English grammar and composition courses do not address many of the unique writing needs and requirements that go into a viable police narrative. Officers may come away from an English class with what, for them, may be useless terminology; overly involved, lengthy sentences; and, in short, a feeling that they have wasted their time. What now?

It was with these concerns in mind that the authors approached the writing of this text. They combined their talents and proven expertise in English instruction and police training to provide students and teachers with a reliable, relevant, and highly readable text. The result is an English text with law enforcement professionals in mind.

All the chapters have an introduction, stated objectives, explanatory text, and practice exercises geared specifically to police interests. At the end of each chapter, there are a chapter review, discussion questions, and review exercises, most in the form of police narratives, not isolated, unrelated sentences such as those found in many traditional grammar texts. In all examples, practices, and exercises, "deadwood words" have been replaced with "clear-meaning," everyday words such as those that should be used in clear, concise, well-written police narratives. Because the authors believe it is important to teach more than report writing and English in criminal justice programs and because they believe it is important to maintain equality as part of the educational process, every practice, exercise, or example includes all races and both genders in a variety of activities.

The first five chapters are devoted to presenting English grammar in a straightforward, easy-to-understand manner, using a conversational-style format. The last two chapters are devoted to the police report writing process. They include the types and uses of individual reports, interviewing and note-taking techniques and strategies, the organization of police reports, and proper word usage: in short, the rudiments of well-written police narratives.

A *Student Workbook* at the end of the text includes additional exercises to provide more practice for students who feel they want or need additional reinforcement. In addition, practice scenarios are provided that afford students an opportunity to put into use all the grammatical concepts and report writing techniques they have learned in the preceding chapters.

The *Teacher's Guide and Answer Key* includes answers to the 167 practice examples, 38 discussion questions, and 355 exercise examples. Special tips or suggestions are included at the beginning of each chapter review.

The entire text incorporates English grammar and composition skills with proven effective police report writing techniques and strategies. It emphasizes the importance of correct English composition for accurate police reports. Moreover, the unique combination of the authors' experience and background—an English instructor of more than 20 years with practical experience in police report writing needs and an active deputy sheriff and expert police trainer—guarantees a text that is relevant for today's law enforcement personnel, as well as being a permanent, personal resource and guide for English grammar and report writing techniques.

The authors wish to acknowledge the following reviewers: Richard R. Becker, North Harris College; Donald G. Hanna, Chief of Police—Champaign, IL; Wayne Madole, Broward Community College; Ronald A. Pincomb, New Mexico State University; and Lois A. Wims, Salve Regina University.

PAINLESS POLICE
REPORT WRITING

Chapter Two

SENTENCE ELEMENTS

Introduction

In Chapter One you learned to identify the parts of speech. In th hap r
you will learn how these parts of speech go together to form sent es. e
elements of a sentence go together to make a complete thou; Th   e
elements include subjects, predicates, direct and indirect objec ub   t
complements, and modifiers.

Objectives

At the end of this chapter, you will be able to do the following:

1. Identify complete sentences.
2. Identify and correct fragments and run-on sentences.
3. Identify sentence elements.
4. Identify subjects and predicates.
5. Identify direct and indirect objects.
6. Identify subject complements and modifiers.
7. Identify phrases and clauses.

COMPLETE SENTENCE

A sentence is a group of words containing a subject and a verb an e   ess
ing a complete thought. It is the basic unit in any writing. The a fou

basic types of sentences: simple, compound, complex, and compound–complex.

Simple Sentence

A simple sentence contains only one main (independent) clause. A clause is a group of words that contains a subject and a predicate.

EXAMPLE:

The officer wrote the citation.

Compound Sentence

A compound sentence contains two or more main clauses, but no subordinate (dependent) clauses.

EXAMPLE:

You may not catch the suspect, but you should try.

Complex Sentence

A complex sentence has one main clause and one or more subordinate clauses.

EXAMPLE:

If the suspect hits tonight, we'll be ready.

Compound–Complex Sentence

A compound–complex sentence has two or more main clauses and one or more subordinate clauses.

EXAMPLE:

If the suspect hits tonight, you go in front, and I'll cover the rear.

FRAGMENT

A fragment is a group of words that is punctuated as a sentence but does not contain a main clause (complete thought). Frequently, a fragment is incorrectly punctuated as a sentence.

EXAMPLE:

Knowing I'd find narcotics. (incorrect)
I searched the house, knowing I'd find narcotics. (correct)

RUN-ON SENTENCE

A run-on sentence is also referred to as a "fused" sentence. A run-on sentence combines two or more independent clauses without punctuation or a coordinating conjunction.

14.   The suspect ran down e street carrying Gold's money.

   _____

   _____

15.   The witness said he'd ca   d the police yelling to Gold.

   _____

   _____

Revise the following sentences   they use parallel c nstructions.

16.   The police arrived, question Gold, and were eginning to search the area.

   _____

   _____

17.   Gold described the suspect as eing tall, we ing black clothes, with brown
   curly hair, and dark eyes.

   _____

   _____

Rewrite the following sentences and c rect all ammatical errors.

18.   The witness was also able to desc e the spect the same as Gold is describ-
   ing except the witness wasn't clos nou to see the suspect's eyes.

   _____

   _____

19.   Officer Carver seen the suspect run   through the park still carrying the
   gun.

   _____

   _____

20.   Officer Frick and his partner Officer   caught the suspect at the corner of
   Fifth and Main.

   _____

   _____

Chapter Three

PRONOUNS

Introduction

Pronouns, one of the eight parts of speech mentioned in Chapter One, are words used in place of nouns. Generally, pronouns are used to refer to nouns that have already been used or implied. These nouns or noun phrases, called *antecedents*, appear before or shortly after the pronouns. Using pronouns shortens your writing and makes it less tedious and repetitious.

Objectives

At the end of this chapter you will be able to do the following:

1. Identify and use pronouns correctly.
2. Write and/or correct sentences using clear antecedents.
3. Write and/or correct sentences containing pronouns with regard to person, number, gender, and case.

TYPES OF PRONOUNS

Pronouns can be divided into several main classes.

Personal Pronouns

Personal pronouns can be either singular or plural.

EXAMPLES:

Either the boys *or* their father was guilty.
Neither the man *nor* his sons were hurt.

A noun or pronoun followed by a prepositional phrase is treated as if the prepositional phrase were not there.

EXAMPLES:

Three *suspects* in the holdup ran to *their* car.
The *book* of matches had no identification on *its* cover.

Case

Pronouns must agree with their antecedents in case. If a pronoun replaces a subject noun, the pronoun must be in the subjective case.

EXAMPLES:

Deputy Sparling received a letter of commendation.
She saved several children's lives.

If a pronoun replaces an object noun, such as a direct or indirect object or an object of a preposition, the pronoun must be in the objective case.

EXAMPLES:

The teller gave *Sgt. Kim* a description of the suspect.
The teller gave *him* a description of the suspect.

When a noun and a pronoun are used together or when two pronouns are used together, you may find yourself trying to decide whether to use the subjective or objective case pronoun. Simply read the sentence with each pronoun by itself; you should be able to tell by the sound of the sentence which pronoun form is correct.

EXAMPLES:

The victim gave Sgt. Seaver and *I or me* the suspect's description
The victim gave *me* the suspect's description.

Gender

Pronouns should agree with their antecedent in gender: masculine femi-
nine. Not long ago it was acceptable to say, "A police officer should consider all *his* options before *he* acts." Today, however, many people object sentences that imply that all officers are male, which is not the case. To liminate
sexism in your writing, you can edit your sentences in several w s.

You can use the double-pronoun construction, which is widely accepted today.

EXAMPLE:

A police officer should consider all his or her options before he or she acts.

As you can see, if repeated, double pronouns become tedious.
Another option is to make both the antecedent and the pronoun plural.

EXAMPLE:

Police officers should consider all *their* options before *they* act.

A third option is to eliminate all reference to specific gender. This is a particularly good option when you can use it.

EXAMPLE:

A police officer should consider all options before acting.

Person

Pronouns in the same sentence must agree in person with each other.

EXAMPLES:

One can live happily in an area if *you* feel *you* have good protection. (incorrect)

One can live happily in an area if *he/she* feels *he/she* has good protection. (correct)

You can live happily in an area if *you* feel *you* have good protection. (correct)

PRONOUN REFERENCE

Pronouns must not only agree with their antecedents, but those antecedents must also be clearly recognizable or referred to. Most problems that occur with pronouns can be traced back to unclear pronoun references: which antecedent is the pronoun referring to or replacing?

EXAMPLE:

If the police dog won't eat its food, try covering it with warm gravy.

The pronoun *it* has no clear antecedent; it could refer to either the dog or the food. Logically, you would want *it* to refer to the food.

Correction:

If the police dog won't eat its food, try covering the food with warm gravy.

Sometimes the only way to correct an unclear pronoun reference is   repeat
the antecedent and eliminate the pronoun.

Practice

Choose the correct pronouns in the following sentences:

> Officer Pickens said him/he and his canine partner worke   well to-
> gether.
> The Chief left the discussion up to we/us officers.
> The store owner gave the description to Officer Stein and I/   e.
> The new position will be offered to either you or I/me.
> Each of the suspects wanted to answer the charges for thems   /es/her-
> self.

CHAPTER REVIEW

Pronouns are useful words that take the place of nouns. The no   s or noun
phrases that pronouns replace are called antecedents. Prond   is can be
classified under several categories. Pronouns must have clear   itecedents
and must agree with these antecedents in person, number, genc   , and case.

DISCUSSION QUESTIONS

1. What is a pronoun?
2. What is an antecedent?
3. List the main types of pronouns and give examples of each ty
4. Pronouns must agree with their antecedents in what four wa

EXERCISES

Underline the pronouns in the following sentences:

1. The officer asked, "Who called the police?"
2. The victim said, "I called. Someone stole my car."
3. The officer asked, "Could someone you know have taken it
4. "I told my family whoever needed it could take the car."

Select the correct pronoun in the following sentences:

5. Do you think one of *they/them* took it?

6. Neither of the girls has *her/their* license.
7. My wife and *me/myself/I* don't think the girls took it.

Circle and correct the pronoun problems in the following sentences:

8. The officer told the victim he didn't have enough information.

   _____

   _____

9. The officer told the victim if he intends to file a report, you'll have to be more helpful.

   _____

   _____

10. He said if he got more information he could call later.

    _____

    _____

Chapter Four

MECHANICS

Introduction

The mechanics of English grammar include capitalization, punctuation, and writing numbers. While these mechanics may not seem important to you in writing police reports, they are. Correct capitalization shows where sentences begin and identifies proper nouns. Punctuation includes the correct usage of periods, commas, colons, and the like. Correctly written numbers eliminate errors.

Objectives

At the end of this chapter, you will be able to do the following:

1. Use capital letters correctly.
2. Punctuate sentences correctly.
3. Write numbers correctly.

CAPITALIZATION

In the following instances, you would commonly use capital letters in police reports. If you block print your reports, capital letters must be larger than the other letters.

First Words

Capitalize the first word in a sentence, in a direct quotation, and in a line of regular poetry.

EXAMPLES:

He is the new officer.
The sergeant said, "Welcome to the department."
We're all happy you are here.
After work, join us for a beer.

Proper Nouns

The first letter of a proper noun is capitalized. The pronoun *I* is always capitalized.

People, Organizations, and Their Members

EXAMPLES:

Walter James
Tulsa Police Department
Captain Walter James

Places and Geographical Areas

EXAMPLES:

New York City
New England
the South, West, North, East, Midwest, Southwest

Do not capitalize directions.

EXAMPLE:

The speeding vehicle turned *east* on Main Street.

Rivers, Lakes, and Mountains

EXAMPLES:

Rio Grande River
Lake Superior
Mt. Rushmore

Ships, Airplanes, Trains, and Space Vehicles

EXAMPLES:

Queen Mary
Concorde

Union Pacific
Discovery

Nationalities, Races, Tribes, Languages, Religions, and Political Parties

EXAMPLES:

Canadian
Caucasian
Navajo
Spanish
Methodist
Democratic

Family Relationships: Capitalize words that indicate family relationships when they are used in place of a person's name.

EXAMPLES:

We asked Dad if we could go.
We asked our dad if we could go.

Titles

EXAMPLES:

General Eisenhower
Dr. Stanley Fine, M.D.

LEARNING
TIP

President, Presidential, Presidency, and Executive are capitalized when they refer to the office of the President of the United States. The same format is followed for the Vice-president.

Deity and the Bible: Capitalize Bible and Biblical when they refer to scripture.

EXAMPLES:

God, Jehovah, Allah

Brand Names and Registered Trademarks

EXAMPLES:

Levis
Coors

Books, Publications, Magazines, Newspapers, Poems, Articles, Headings of Chapters, Plays, Television Shows, Songs, Paintings, and Other Works of Art: Capitalize the first word and all important words in titles of all the examples listed above.

EXAMPLES:

The Choirboys
Time, Los Angeles Times
Miracle on 34th Street

Days, Months, Holidays, and Holy Days

EXAMPLES:

Thursday, November 25th is Thanksgiving.

Courts

EXAMPLE:

the Supreme Court

Historical Events, Documents, and Time Periods

EXAMPLES:

the Great Depression
the Bill of Rights
the Middle Ages

Adjectives Formed from Proper Nouns: You capitalize the adjective, but not the noun that follows it.

EXAMPLES:

German shepherd
French restaurant

Names of the Seasons: Capitalize them only when they refer to specific seasons of specific years.

EXAMPLES:

The Winter of 1840 was the worst on record.
This winter has been very mild.

Correspondence: You capitalize the salutation and the first word in the complimentary closing of a letter.

EXAMPLES:

Dear Carl,
Sincerely yours,

Practice

Correct all capitalization errors in the following sentences:

sergeant Davis said, "merry christmas."
On Christmas Day, the Sergeant said, "Merry Christmas."

EXAMPLE:

$445.78

You spell out indefinite sums of money.

EXAMPLE:

The suspect took approximately fifty dollars.

Practice

Select the correct usage of numbers in the following:

3/Three victims came to the station.

The youngest was 21/twenty-one years old.

One victim was missing exactly $23/twenty-three dollars.

The other was missing approximately $30/thirty dollars.

The police car took 18/eighteen gals. of gasoline.

CHAPTER REVIEW

You have learned the mechanics of English grammar, including capitalization, punctuation, and writing numbers. Correct capitalization shows where sentences begin and identifies proper nouns. Punctuation includes correct usage of periods, commas, colons, and the like. Correctly written numbers eliminate errors.

DISCUSSION QUESTIONS

1. When do you use capital letters?
2. What are the end punctuation marks and when are they used?
3. Fifteen uses of commas were discussed in the chapter. Name five and explain them.

4. Explain the usage of semicolons and give one example.
5. Explain the usage of colons and give two examples.
6. Explain the three uses of apostrophes and give examples of each.
7. Explain apostrophe usage in forming possessives and give examples of each.
8. Explain how hyphens are used to divide words.
9. Define the term ellipsis and explain why caution is necessary in its use.
10. When do you write out numbers and when do you use numerals?

EXERCISES

Correct the capitalization and number errors in the following sentences. Add correct punctuation where necessary.

1. Mrs Wood called the police department to report a Burglary.
2. She said, I've been robbed
3. Officer allen arrived at 230 to take the report
4. He asked her, "What happened"
5. She told him the following
6. She left the house at 9:00 *A.M.* and went to the store then to the gas station and finally to the Cleaners.
7. when she got home she found the front door open.
8. When she went inside the house had been ransacked the living room furniture was overturned.
9. She said the following items were missing 1 a copy of Gone with the Wind 2 her twenty one year old daughters address book and 3 one gal. of wine.
10. She wasnt sure, but thought the total loss was approximately $450.

Chapter Five

SPELLING

Introduction

Accurate spelling is essential in police report w   g. Your spelling errors can confuse the reader and change the meani   f your report. Spelling errors can also make you look less competent t   n you are. In this chapter you will learn some of the rules of spelling   help you correctly spell common words that may cause you trouble.

Objectives

At the end of this chapter, you will be able to   the following:

1. Spell words correctly.
2. Add suffixes and prefixes correctly.
3. Form plurals correctly.
4. Identify seven resources to aid you i orrect spelling.

RULES OF SPELLING

Since there are many exceptions in Englisł   ing, no one set of spelling rules will cover all cases. When you are in d   use a dictionary. However, the following rules will help you in many   ns.

ie or ei

When the sound is long *ee,* use *i* before *e,* except after *c.* Remember the old folk rhyme that says:

Use *i* before *e*   (e.g., believe, field, piece)
Except after *c*   (e.g., receive, ceiling, deceive)
Or when sounded like *a*
As in *neighbor* and *weigh.*   (e.g., freight, reign)

> There are 10 exceptions to the above rule. It would be helpful for you to memorize them.
>
> *Exceptions:* either, neither, leisure, seize, weird, foreign, height, counterfeit, forfeit, science

Prefixes

A prefix consists of one or more letters added before the root word to make a new word. You do not have to worry about single or double letters; simply write the prefix and add the root word as it is normally spelled.

EXAMPLES:

Prefix	Root Word	New Word
un	known	unknown
un	necessary	unnecessary
mis	spelled	misspelled

Suffixes

Suffixes are one or more syllables added after the root word. Unfortunately, adding a suffix is not as simple as adding a prefix.

Final Silent *e*: When adding a suffix to the word, you drop the final silent *e* if the suffix begins with a vowel (*a, e, i, o, u*).

EXAMPLES:

Root Word	Suffix	New Word
come	ing	coming
imagine	ary	imaginary
berate	ed	berated

The word's last two letters must be one vowel and one consonant.

EXAMPLES:

commit + ed = committed, stop + ing = stopping

When adding *-ly* to a word ending in *-l*, keep the final *-l*.

EXAMPLES:

careful + ly = carefully, brutal + ly = brutally

When adding *-ness* to a root word ending in *-n*, keep the final *-n*.

EXAMPLES:

open + ness = openness, green + ness = greenness

Not Double Consonants: When the following conditions are present, you do not double the final consonant:

When the word ends in two or more consonants.

EXAMPLES:

jump + ed = jumped, find + ing = finding

When two or more vowels precede the final consonant.

EXAMPLES:

contain + ing = containing, appear + ed = appeared

When the word ends in a single accented vowel and a consonant and the suffix begins with a consonant.

EXAMPLES:

regret + ful = regretful, equip + ment = equipment

If the accent is not on the last syllable of the root word.

EXAMPLES:

quarrel + ing = quarreling, bigot + ed = bigoted

Plurals

You make singular nouns plural using several different spelling guidelines.

Adding -s: You add *-s* to make most nouns plural.

EXAMPLES:

books, automobiles, guns, suspects

Adding -es: You add *-es* to nouns ending in *s, ch, sh, x,* or *z*.

EXAMPLES:

Joneses, boxes, flashes, churches, inches, buzzes

Words ending in o: When words end in the letter *o*, use *-s* if a vowel comes before the final *-o*. If a consonant comes before the final *o*, use *-es*.

EXAMPLES:

radios, scenarios, studios

heroes, potatoes, vetoes

There are four exceptions: memos, pros, pianos, solos.

Words Ending in f or fe: For some words ending in *f* or *fe*, you change the *f* to *v* and add *-s* or *-es*. Some words don't change the *f* to *v*; just add *-s*.

EXAMPLES

knife/knives, wife/wives, chief/chiefs, staff/staffs

Irregular Plurals: Some nouns have irregular plurals.

EXAMPLES:

foot/feet, child/children, man/men, woman/women

Unchanged Nouns: Some nouns do not change when you form the plural.

EXAMPLES:

sheep, moose, series, species

Compound Words: You make the last part of the compound word plural when the word is not hyphenated or written as two separate words. When they are written as two separate words or hyphenated, you make the most important part of the word plural.

EXAMPLES:

briefcases, mailboxes, brothers-in-law, bus stops

Simple plural forms *never* have an apostrophe (') before the *-s* ending.

EXAMPLES:

Two boys left. (correct)

Two boy's left. (incorrect)

techniques
testimony
than
their
then
there
they're
thieves
thorough
to/too/two
traffic
transferred

trespassing
truancy
unnecessary
vagrancy
victim
villain
warrant
woman
write
writing
written

HOMONYMS

Homonyms are words that sound alike, but have different meanings and are spelled differently. You should learn the difference and use the correctly spelled word.

beat/beet
boar/bore
board/bored
bread/bred
break/brake
bridal/bridle
buy/by/bye
capital/capitol
ceiling/sealing
cent/sent/scent
cereal/serial
cite/sight/site
chord/cord
corps/corpse
council/counsel/consul
current/currant
dear/deer
hole/whole
idle/idol
its/it's
knew/new
knot/not
know/no
liable/libel

lain/lane
lessen/lesson
lie/lye
loan/lone
made/maid
maybe/may be
meat/meet
medal/meddle
muscle/mussel
naval/navel
oar/or/ore
ordinance/ordnance
pail/pale
pain/pane
pair/pare/pear
pause/paws
peace/piece
peal/peel
pedal/peddle
peer/pier
plain/plane
pray/prey
presence/presents
pride/pried

principal/principle
rain/reign/rein
raise/rays/raze
rap/wrap
real/reel
right/rite/write
road/rode/rowed
role/roll
sail/sale
scene/seen
seam/seem
sense/cents
serf/surf
shear/sheer
shone/shown
soar/sore
sole/soul
stairs/stares
stake/steak
stationary/stationery
steal/steel

straight/strait
tail/tale
taught/taut
team/teem
tear/tier
their/there/they're
throne/thrown
through/threw
tied/tide
to/too/two
toe/tow
vain/vane/vein
vale/veil
vial/vile
wail/whale
waist/waste
wait/weight
waived/waved
way/weigh
weak/week
wear/where

ABBREVIATIONS

Three Rules for Abbreviations

1. Spell out all titles except Mr., Mrs., Mmes., Dr., and St. (saint, not street).
2. Spell out street, Road, Park, Company, and similar words used as part of a proper name or title.
3. Spell out Christian names (William, not Wm.).

Standard Abbreviations

You should exercise caution when using abbreviations. While they may shorten reports, using incorrect or unfamiliar abbreviations can lead to misunderstandings. When in doubt, spell it out.

Dates, Time, and Measurement

Jan.	Apr.	Jly.	Oct.
Feb.	May	Aug.	Nov.
March	June	Sept.	Dec.

Mon.	Thurs.	Sun.
Tues.	Fri.	
Wed.	Sat.	

1st	6th
2nd	7th
3rd	8th
4th	9th
5th	10th

yr.	year	in.	inch
yrs.	years	ft.	feet/foot
mo.	month	yd.	yard
mos.	months	mi.	mile
wk.	week	g.	gram
wks.	weeks	kg.	kilogram
hr.	hour	km.	kilometer
hrs.	hours	lb.	pound
min.	minute	lbs.	pounds
mins.	minutes	oz.	ounce
sec.	second	meas.	measurement
secs.	seconds	doz.	dozen
		l.	length
		wt.	weight
		hgt.	height
		w.	width

Common Abbreviations

administration	admin.
all points bulletin	APB
also known as	AKA
amount	amt.
approximate	approx.
assistant	asst.
assist outside agency	AOA
attempt	att.
Attempt to locate	ATL
attorney	atty.

be on the lookout	BOLO
birthplace	BPL
building	bldg.
burglary	burg.
captain	capt.
caucasian	cauc.
central	cen.
chief of police	COP.
Colonel	Col.
Company	Co.
convertible	cvt.
court	ct.
date of birth	DOB
dead on arrival	DOA
defendant	def.
degree	deg.
department	dept.
Department of Motor Vehicles	DMV
detective	det.
description	descp.
director	dir.
district	dist.
division	div.
driver's license	DL
Doctor	Dr.
doing business as	DBA
driving under the influence	DUI
driving while intoxicated	DWI
eastbound	E/B
enclosure	encl.
example	ex.
executive	exec.
federal	fed.
general broadcast	GB
government	govt.
headquarters	hdq.
highway	hwy.
hospital	hosp.
identification	ID
informant	inf.
inspector	insp.
junction	junc.
junior	jr.

juvenile	juv.
last known address	LKA
left	L
left front	LF
left hand	LH
left rear	LR
license	lic.
Lieutenant	Lt.
Lieutenant Colonel	Lt. Col.
location of birth	LOB
Major	Maj.
manager	mgr.
maximum	max.
medium	med.
memorandum	memo
middle initial	MI
misdemeanor	misd.
modus operandi	MO
National Auto Theft Bureau	NATB
National Crime Information Center	NCIC
no further description	NFD
no middle name	NMN
northbound	N/B
not applicable	NA
number	no.
numbers	nos.
officer/official	ofc.
Ohio driver's license (NMDL, FDL, etc.)	ODL
opposite	opp.
organization	org.
package	pkg.
page	p.
pages	pp.
passenger	pass.
permanent/personal identification number	PIN
pieces	pcs.
pint	pt.
place	Pl.
place/point of entry	POE
point of impact	POI
police officer/probation officer	PO
quantity	qty.
quart	qt.

received	recd.
required/requisition	req.
right	R
right front	RF
right rear/rural route/railroad	RR
road	Rd.
school	sch.
section	sect.
Sergeant	Sgt.
serial	ser.
southbound	S/B
subject	subj.
Superintendent	Supt.
surface	sur.
symbol	sym.
tablespoon	tbsp.
technical	tech.
teletype	TT
transportation	tran.
treasurer	Treas.
University	Univ.
unknown	unk.
vehicle identification number	VIN
veterinarian/veteran	vet.
village	vil.
volume	vol.
weapon	wpn.
wholesale	whsle.

TIPS

The following seven tips should help you improve your spelling.

Speller's Journal: When you get a report back with misspelled words, write those words (correctly spelled) in the back of your notebook. We use the same words over and over, and soon you will correctly spell those words.

Dictionary: When in doubt, always use a dictionary to verify spelling and meaning.

Speller/Divider: Speller/dividers are pocket-sized books that list the words correctly spelled. They don't have any definitions. Most officers know what a word means, but may not know how to spell it.

Thesaurus: A thesaurus is a book of synonyms, words with similar meanings. Use by police officers improves spelling and makes reports more interesting. It also helps you find a word with the exact meaning you need.

Misspeller's Dictionary: If you have trouble finding correctly spelled words in a dictionary, try a *Misspeller's Dictionary*. The words are listed incorrectly spelled, followed by the correct spelling, for example, *newmonia/pneumonia*.

Electronic Spellers: Hand-held, battery-powered electronic spellers are available. The more sophisticated models include a dictionary and thesaurus. They will correctly spell a word, but may not differentiate between homonyms.

Proofreading: Proofread your own work or have someone read it for you. Proofreading will greatly reduce spelling errors.

CHAPTER REVIEW

You have learned that accurate spelling is essential in police report writing. You learned some of the rules of spelling to help you correctly spell common words that may cause you trouble. You also learned the importance of correctly using homonyms and abbreviations

DISCUSSION QUESTIONS

1. What is a homonym?
2. How can homonyms cause trouble in your reports?
3. What are some of the resources you can use to check your spelling?

EXERCISES

Add the suffixes *-ed* and *-ing* to the following words:

raid	_____
prevent	_____
describe	_____
rob	_____
try	_____
identify	_____
study	_____
die	_____
allege	_____
hope	_____

Add *-able* to the following words:

rely   _____

note   _____

remark   _____

work   _____

excite   _____

Write the plural form of the following:

officer   _____

witness   _____

child   _____

man   _____

attorney   _____

knife   _____

party   _____

box   _____

radio   _____

shelf   _____

Circle the correctly spelled words from the choices you are given:

accellerated	excelerated	accelerated
all right	allright	alright
burglary	burglery	berglary
dialated	dilated	diliated
homocide	homicide	homacide
lisence	license	licence
preceeded	preceded	precceeded
sargent	sergent	sergeant
secratery	secretary	secretery
warrent	warrant	warent

Circle the correct homonym.

1. If the suspect walks down that (aisle, isle, I'll), I'll meet her at the front of the theater.

2. Reporters should not be (allowed, aloud) to leak information before a trial.

3. Smith refused to pay the (fair, fare) because she said the amount wasn't (fair, fare).

4. Officers found the jewel-encrusted (idle, idol) in the suspect's closet.

5. The (knew, new) recruit said he (knew, new) some of the local ordinances.

6. The witness was very (pail, pale).

7. He (passed, past, pasted) the detour because he drove right (passed, past, pasted) it.

8. This time, the detective obtained the (right, writ, write) (right, writ, write) so the judge didn't have to (right, writ, write) another.

9. The defense attorneys were unable to save (their, there, they're) client because (their, there, they're) wasn't evidence to refute the charges.

10. She said the (vial, vile) man had thrown the (vial, vile) of acid at her.

Chapter Six

POLICE REPORTS

Introduction

The previous chapters provided you with English composition skills. In this chapter you will learn about the different types of police reports, how they are used, and what makes a good report. Remember, the television and movie glamour of police work seldom includes report writing.

Objectives

At the end of this chapter, you will be able to do the following:

1. Define what a report is.
2. Identify the types of reports.
3. Define the uses and purposes of reports.
4. Define the qualities of a good report.

DEFINITION OF A REPORT

Traditionally, a report meant a "police report," or the narrative you have to write after completing an investigation. But, actually, reports take many different forms. A report is defined as the following: any documentation recorded on a departmental form, or other approved medium (computer disks), and maintained as a permanent record.

Newspapers and the Media

Crime reports, and in some cases all reports, are available to the press and media. In most states, some parts of the crime reports may be deleted, for example, names of juvenile suspects and victims and the victims of certain crimes. However, in general, the press has the right of access to reports. The result is they read exactly what you have written, including misspellings and grammatical errors.

Reference Material

Because reports are permanent documents, they provide an excellent source of historical information. They may be used to document the agency's actions, refresh your memory, or determine liability.

Statistical Data for Crime Analysis

The rapid development of computer technology, including expert systems and automated pin maps, has resulted in improved crime analysis. The source document for that information remains the crime report you write in the field. Your reports are used to identify trends, locations, and methods of operations. The result of that analysis may be directed patrol.

Documentation

Reports are used to document the action of the department and its officers. Because police departments are typically reactive, reports document what actions were taken to stop the criminal activity or arrest the suspect. They provide evidence of the department's responsiveness to the community and its needs.

Officer Evaluation

It is common for supervisors to use reports to evaluate an officer's performance. An experienced supervisor can determine your ability to organize information, level of education, technical knowledge, intelligence, and pride in the job. A report discloses an officer's weaknesses, weaknesses the officer may not even realize he or she has.

Statistical Reporting

Crime reports are the source document for the collection of statistical data. Agencies report crime statistics to various state and federal agencies. Statistical reports may also be generated for budget purposes, city council briefings, and other special-interest groups.

Report Writing Audience

Your reports must be self-explanatory because numerous people make decisions based on the information in your reports. Depending on the nature of the event, any or all of the following may read your report:

Police departments: Supervisors and administrators of both your department and those cooperating in investigations

Attorneys: Prosecution, defense, civil, and judges—all attorneys who may read your reports

Jurors: In both criminal and civil trials

Administrators: From your department and jurisdiction, as well as from city, county, and state jurisdictions

Medical professionals: Doctors, psychiatrists, and psychologists

Corrections: County jail and state and federal prison staff, including probation officers and parole agents

Insurance companies: The parties involved in claims

Media: Newspapers, radio, and television

Regulatory agencies: Motor vehicle departments, insurance commissioners, alcohol beverage control, consumer affairs

If any part of your report requires further explanation, you have failed to accomplish your objectives. When you have to write a supplemental report to explain your original report, you create an air of skepticism. Your credibility may become questionable in the eyes of the court. You must not evade the necessity of well-written reports. It is important that you understand the merits of effective report writing and recognize the significance of reports in the total criminal justice system.

QUALITIES OF A GOOD REPORT

All police reports must contain certain qualities, which can be categorized as *accurate, clear, complete, concise, factual, objective, and prompt.*

Accurate

Accurate means in exact conformity to fact: errorless. A fact is something that has been objectively verified. You must report the facts correctly and without error. The identification of facts is imperative to establish the *corpus delicti* or body of the crime in your report. You must restrict your report to the facts of the incident as you saw them or as victims and witnesses reported to you. You must accurately report the conditions of the scene as you found them. Avoid reporting opinions, inferences (drawing a conclusion), suppositions (assumption of truth), or hearsay as though they were facts.

Clear

The language and format in your police report must be simple and to the point. Clear means plain or evident to the mind of the reader. You should use simple words so the reader will know exactly what you want him to know. Avoid using words that can have double meanings, slang, jargon, and unnecessary abbreviations. Use active voice, past tense, first person sentences to answer:

Diagrams and photocopies make your reports more effective.

Diagrams can convey facts even more accurately and briefly than a narrative explanation. Photocopies of some items, Miranda cards, consent search cards, written notes or statements, and the like, may be attached to your reports. Other items, such as weapons, tools, or contraband, could also be copied and attached to reports. A copy of a three-dimensional object is worth a page of narrative and affects the reader more.

CHAPTER REVIEW

In this chapter you learned what the definition of a report is and that there are 10 different types of police reports. You also learned about the uses and purposes of police reports, how they are used, who uses them, and what makes a good report.

DISCUSSION QUESTIONS

1. What are the ten types of reports?
2. What are seven uses and purposes of reports?
3. Who makes up the report-writer's audience?
4. What are the seven qualities of a good report?

Chapter Seven

REPORT WRITING TECHNIQUES

Introduction

The previous chapters provided you with English composition skills and general knowledge about the types of police reports. In this chapter you will learn to apply your skills and knowledge to write police reports. Police report writing is considered technical writing, and as such you will need to develop special skills and techniques. Police report writing is the backbone of criminal investigations and prosecutions.

Objectives

At the end of this chapter, you will be able to do the following:

1. Define and explain interpersonal communications.
2. Identify the five parts of the report writing process.
3. Define chronological order.
4. Identify and write active-voice sentences.
5. Identify appropriate word usage for police reports.
6. Identify the advantages of first versus third person.
7. Properly use a tape recorder for note taking and report dictation.

INTERPERSONAL COMMUNICATIONS

You must understand the interpersonal communication process before you learn to conduct interviews and interrogations. One of your most valuable tools as a police officer is your interpersonal communication skill.

Definition of Communications

Generally, communication is defined as the use of language, spoken or written, to exchange ideas or transfer information. The transfer of information or ideas from one person to another includes the transmission and receipt of a message to effect some type of action or change.

Reasons for Communication

There are four reasons you communicate with other people.

1. *To provide adequate information for group living:* Police services are delivered to multicultural communities that include a growing elderly population. Your role as a police officer is rapidly changing from the traditional enforcer of laws to that of a service provider.
2. *To clarify perceptions and expectations:* The exchange of ideas and information is essential to clarify your perceptions and expectations and those of the community you serve.
3. *To stimulate creative thinking through feedback:* The human mind requires stimulation. You receive that stimulation from the feedback you receive during the communication process.
4. *To maintain your balance in the world:* During the communication process, you receive reinforcement or reassurance that you are okay.

Communication Process

The communication process contains a *sender*, *receiver*, and the *feedback loop*. There is a continuous line of communication between the *sender* and *receiver*. They are linked together by the *feedback loop*. When you begin an interview, you are the *sender* because you ask questions. The person you are interviewing is the *receiver*. Both of you listen to and watch each other, which provides you *feedback*. When the person you're interviewing answers, your roles in the communication process change. Feedback includes the answers to questions, gestures, and expressions.

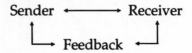

Types of Communication

You use verbal and nonverbal communication when dealing with people. Remember that you not only give off communication clues, but you should also practice reading the clues others give you.

Nonverbal Communication: You use three types of nonverbal communication: gestures, facial expressions, and body language.

Your gestures should be nonthreatening, using an open hand, for example. Facial expressions reveal your true feelings, so master appropriate expressions for every situation. Body language is easily read by others and conveys both your feelings and meanings. Positive use of body language will encourage people to talk and improve your ability to gather information.

Verbal Communication: You may not realize it, but there are many different types of verbal communication. It is important for you to recognize and understand each type.

One-way communication is lecture or direction. It is effective in limited situations, especially where compliance without feedback is necessary. An example would be felony or high-risk car stops.

Two-way communication includes speaking and listening. You typically exchange information or ideas in two-way communication.

Oral-in-person communication means you can see and hear the other person you are speaking to. You have the opportunity to use both verbal and nonverbal communication skills.

Oral-telephone communication is just what it sounds like, talking on the phone. The disadvantage is nonverbal communication is eliminated.

Written communication is the most difficult to master. You always disclose something about yourself in your writing. Typically, you disclose your ability to organize information, your educational level, and your technical expertise.

Practice

Team up with another student in the class and alternate between each other, doing the following:

> Look at your partner and use your eyes and facial expressions to convey warmth and caring.

> Look at your partner and use gestures to convey that "you mean business."

> Look at your partner, smile, and say, "I'm going to fire you."

> Look at your partner, smile, and say, "I'm always here to help you."

REPORT WRITING PROCESS

Police report writing is a five-step process. They are (1) interviewing, (2) note taking, (3) organizing and planning, (4) writing the narrative, and (5) proof-reading. Your preliminary investigation includes report writing. During the investigation you must complete each of the steps, or your final report will not be acceptable.

Interviewing

Interviewing is the first step in the process. Your interviews of victims, witnesses, and suspects are the backbone of your preliminary investigation. Frequently, the first officer at the scene of a crime has the best opportunity to solve the crime by conducting a thorough preliminary investigation. If you do not conduct successful interviews, your preliminary investigation and police report will not be acceptable.

Three-step Interview Method: The three-step interview method is an easy way for you to conduct interviews. It is structured to allow your informant to speak at ease while you have control over the interview. While you are learning, practice on simple interview situations. After you feel comfortable, you will be able to use the three-step method for interviews and interrogations.

1. *Subject tells the story:* You listen, keeping the subject on track, and giving verbal and nonverbal clues to keep the story flowing. You don't attempt to take notes during this first step. During this step, you accomplish the following:

   Establish rapport with the subject.
   Determine what crime, if any, occurred.
   Determine what agency has jurisdiction.
   Detect signs of untruthfulness or discrepancies.
   Determine what action you should take.

2. *Subject retells the story and you take notes:* You gather identifying information about the subject and ask questions about the incident

as you guide the subject through the story a second time. During this step, you accomplish the following:

Determine the chronological order of the event.

Establish the *corpus delicti* of the crime.

Ask questions in the order you want to write the report (thereby organizing your report as you take notes).

3. *You read your notes to the subject:* When you do this, you are actually writing your report for the first time. What you are reading is going to be what you write in your report. During this step, the subject can correct errors or remember additional information, and you can correct errors or ask additional questions.

> Use of the three-step interview method adds credibility to your court-room testimony. You can testify that you always use the same method, following the same three steps, in all the interviews you conduct. If there is a discrepancy in the informant's courtroom testimony and what you've written in your report, your use of the three-step method documents your actions.

Interrogation: In general, there is one difference between an interview of a subject and the interrogation of a suspect: *focus.* An interrogation is a planned interview with the primary focus being to obtain a confession or eliminate the person as a suspect. A secondary focus may be to find incriminating evidence. Generally, when you interview a victim or witness, you are not sure what he or she is going to tell you. When you interrogate a suspect, you have probably already collected substantial information about the incident and have a pretty good idea what the suspect is going to say. During most interrogations, police officers don't ask questions they don't know the answers to. That doesn't mean you will get the answer you anticipated.

You use the same interpersonal communication skills in both situations. That's why it is important for you to develop the ability to use both your verbal and nonverbal skills. In some cases you may have several questions written down to ask a suspect during the interrogation. There is nothing in an interrogation to justify coercion, excessive force, or violation of a suspect's constitutional rights. Because case law changes rapidly and may vary from state to state, you should review legal updates and department training bulletins regularly.

Note Taking

Notes are defined as brief notations concerning specific events that are recorded while fresh in your mind and used to prepare a report.

Types of Notes: There are two types of notes: *permanent* and *temporary.* If you use permanent notes, you must keep those notes in a safe storage place. If you elect to use temporary notes, you must discard your notes after completing your report. Whichever type of notes you elect to use, you must not change back and forth based on the type of report. If you do change the type of notes you take from call to call, defense attorneys may attack your credibility by asking why you keep some notes and discard others.

It is recommended you use permanent notes. Recording your notes in a note pad or steno pad provides you with a reference and resource book. You will always have access to your original source of information.

Purposes of Notes: Notes are the basis for your report. You take notes to assist your memory with specific details, such as names, dates of birth, serial numbers, addresses, and phone numbers. If you take neat and accurate notes, and they are subpoenaed into court, they add to your credibility and demonstrate your high level of proficiency and professionalism. A good defense attorney may attempt to develop impeachable inconsistencies between your notes and your report.

Scratch Outlinings: You should use a *scratch* or informal outline for taking notes. Scratch outlines provide a simple, structured, easy way to organize the information on your note pad during step 2 of the three-step interview method.

A scratch outline has a key sentence followed by supporting points. The supporting points are written under the key sentence and indented from the left margin. Neither the key sentence nor supporting points are numbered or lettered.

> **EXAMPLE:**
>
> Key sentence   Ofc. Cleveland arrested Rusty Hodges for burglary
> Supporting points   victim saw Hodges
>   Hodges ran from Ofc. Cleveland
>   Cleveland caught Hodges
>   Hodges had the victim's license

Scratch outlines don't have a set format. Use your own style and develop something useful for you.

The questions you ask in step 2 of the interview are generally your key sentences. The informant's responses are your supporting points. Don't try to write down everything the subject tells you, unless you're going to quote him. Remember, notes are *brief notations*, just enough to remind you at a later time when you are writing your report.

Scratch outlining may sound familiar to you. Teachers use a similar method to teach writing and paragraphing. A key sentence is the same as a topic sentence. The difference between police report writing and English compositions is the source of information and the method used to collect it. Most of the information in a police report comes from interviews you conduct.

When you are taking notes, remember to leave spaces between the lines and don't try to use every inch of the page. You may need to add additional information or correct something you have already written down. You will also need space when you begin to organize your report.

Organizing and Planning

The organizing and planning of your report are the third step. If your report is properly organized and planned, it will be clear, easy to read, understandable, and concise. The small amount of time you spend on organizing and planning greatly reduces the time you spend rewriting reports.

Organizing and planning your report begins during the three-step interview method. During step 2, ask questions in the order you want to include the information in your narrative. Organizing and planning your narrative are closely related to chronological order, which is discussed later in this chapter. Ask questions in the order the event took place, which is the chronological order of the event, and it will make writing your narrative much easier.

Once the informant has told you his or her story during step 1 of the three-step interview, you can begin organizing and planning your report. You will know the chronological order of the event, so ask questions in that order during step 2 of the interview.

Review your scratch outline and verify the information you will need to include in the narrative. You may want to underline, number, or otherwise mark important points in your notes. In some cases you may even draw arrows to remind you where to include information in the narrative. You should also consider the information you want to omit from the narrative.

Writing the Narrative

You may not believe it, but writing the narrative should be the easiest part of report writing. If you have carefully followed the three-step interview method (properly taken notes), and spent a few minutes organizing and planning, writing the narrative is almost anticlimatic. If you use the methods described above, you will be prepared to write reports using the one-write system, dictation, or a lap-top computer.

Just before you begin to write the narrative, stop and think about what you have done and what you have left to do. You have collected all the appropriate information, determined your actions, taken notes, and actually recited the report out loud during step 3 of the interview. Your notes, in scratch outline format, are the road map for writing the narrative. The key sentences generally remind you to start a new paragraph, and the supporting points are used to write the sentences. If you practice following these steps, you will find writing the narrative really is the easiest part.

Proofreading

You may think writing the narrative is the final step, but it's not. When you have finished writing the narrative, proofread it. Most officers are just thankful to have finished the report and don't take the additional moment or two to review their work. Think about who else is going to read the report. Depending on the type of report and whether or not you've arrested a suspect, your report will be read by sergeants, investigators, prosecutors, defense attorneys, and judges. If you have made an arrest, the defense attorney's best chance to defend his or her client comes from your report.

Check for the following when you proofread:

Correct report form(s) and format
Probable cause to stop, detain, arrest, search, and seize
Corpus delicti, the elements of the offense
Correct spelling
Active-voice sentence structure
Proper punctuation

The last things to ask yourself when proofreading are as follows: Is this report the best I can do? Would I want to read it to the chief of police or to a jury? Is there anything else I can do to make it better?

Practice

Team up with another student in the class and use one of the following questions to interview each other. Use the three-step interview method and scratch outlining for note taking.

Tell me everything you've done today, from the time you woke up.

Tell me about your last job.

CHRONOLOGICAL ORDER

You must understand chronological order to write coherent and accurate police reports. Your reports should not skip around or jump from topic to topic. Frequently, officers' reports will jump from interview to interview, which makes the report confusing and difficult to follow. If you use the

three-step interview method and scratch outlining, you shouldn't have any trouble with chronological order.

Definition

For the purposes of police report writing, chronological order is defined as the arrangement of events and/or actions in order by the time of their occurrence. Simply stated: in order, what happened and when.

There are usually two chronological orders to an event: the order of the officer's activities and the order of the event. The exception is when you initiate the activity, for example, an observation arrest, and you become a participant in the event.

Officer's Order of Activity

The order of your activities should be the easiest for you to follow. You will not only have your memory, but also your notes. It is recommended that you write your reports in the order of your activity. This style is frequently called narrative style report writing. (Category style is discussed later in this chapter.)

Think of a residential burglary where the victim calls the police and wants to make a report. What do you do after you receive the radio call?

Interview the victim.

Search for and interview any witnesses.

Search for and possibly arrest the suspect.

Account for your activities.

You would write the narrative in the above example in that same order. First, write about the victim interview, then your search for witnesses, their interview(s), your search for the suspect(s), and so on.

Order of the Event

The order of the event is the arrangement of occurrences and/or actions as they occurred during the crime. When did the suspect enter the house, what did he take, when did he leave, what was he driving, and what was his direction of travel? Every witness may have a different version of the chronological order of the event because they may not have seen the same things.

Using the above example, the following scratch outline demonstrates the chronological order of an event as you might write in your notes.

EXAMPLE:

Victim: left at 7 P.M. to go to the movies

returned at 10:30 P.M.

front door was open

VCR and camcorder were missing from family room

called police

Witness: lives next door at 9380

8:00 P.M. saw suspect walk across victim's yard

EXAMPLE:

I wrote the citation.   (active voice)

The citation was written by me.   (passive voice)

THREE STEPS TO DETERMINE ACTIVE VOICE

Use the following three steps to write in the active voice.

1. Locate the *action* (verb) of the sentence.
2. Locate who or what is doing the action. This is the *doer* (subject) of the sentence. If the *doer* is implied and not written in, or it is being acted on by the *action*, the sentence is weak or passive. If the *doer* is written but not located just in front of the *action*, the sentence is weak.
3. Put the *doer* immediately in front of the *action*.

EXAMPLES:

The officer wrote the citation.   (active voice)

The dispatcher repeated the address.   (active voice)

A suspect was arrested.   (passive voice)

Using the above steps, correct the last statement:

1. What is the *action* of the sentence? *was arrested*
2. Who or what is the *doer* of the action? The sentence doesn't have a *doer*, so who made the arrest? *I*
3. Put the *doer* immediately in front of the *action*: I arrested the suspect.

Practice

Determine if these sentences are active or passive voice. Use an A for active and P or passive.

_____   The sergeant read the crime warning.

_____   The suspect didn't resist arrest.

_____   The meeting was called by the chief.

_____   It was determined by the victim what was missing.

_____   Several citations were written by the motor officer.

OBJECTIVE REPORT WRITING

You learned the importance of objective reports in Chapter Six. While you may write only factual statements in your reports, it is possible they are not objective. In your police reports, objective means you weren't influenced by emotion or prejudice. Your writings are fair, impartial, and not opinionated.

Your reports can lose objectivity because of poor word usage, omission of facts, and uncontrolled personal feelings. You maintain your objectivity by using nonemotional words, including both sides of every story, and remaining a professional during all investigations.

Nonemotional Words

You should use *denotative* words, words that are explicit and nonemotional. Emotional words are called *connotative* because they suggest or imply something beyond the explicit or literal meaning of the word.

EXAMPLES:

bureaucrat, blubbered, scream, wail   (connotative)
public employee, civil servant, cried, wept, yell   (denotative)

Slanting

You slant your report when you fail to include both sides of the event. You must include statements from all the witnesses, victims, suspects, or participants. When you omit all or part of someone's statement, no matter how unusual, you have slanted your report and lost objectivity.

Practice

Circle the denotative word in each example.

woman/broad, officer/cop, confused/crazy, uncooperative/hostile, argue/verbal confrontation

WORD USAGE

Police officers from all regions of the country tend to use similar words and phrases. Unfortunately, these words and phrases are not necessarily the best choices for clarity, objectivity, and conciseness. In most cases,

Instead of:	*Try:*
in accordance with	with, by, as, under
in a most careful manner	carefully
in a number of cases	some, many
in a timely manner	promptly
in connection with	with
in the amount of	for
inasmuch as	because
indebtedness	debt
indicate, state	show, tell, said, noted
initial	first
initiate	begin, start
in large measure	largely
in lieu of	instead of
in many cases	many
in order that	so
in order to	to
in regard to	concerning
in relation to	about
in spite of the fact that	although
institute	start, begin
in the affirmative	yes, agreed
in the course/case of	in, at, or, during, while
in the event of/that	if
in the magnitude of	about
in the majority of cases	usually
in the matter of	in, about
in the time of	during
in the very near future	soon
in the vicinity of	near
in this day and age	today
in view of the fact	since, because
is as follows	follows
it is my understanding that	I understand
it is our opinion	we feel, believe
it should be noted that	furthermore
justification for	reason for
kindly arrange to send	please send
locality	place
locate	find, put
likewise	and also
maintenance	upkeep
make a decision	decide
make a determination	determine
make application for	apply
make inquiry regarding	ask
modification	change
negative results	found nothing
nevertheless	but, however
nonavailability of	unavailable
notwithstanding	despite, in spite of
numerous	many
objective	aim
obligate	bind
obligation	debt
observed	saw
obtain	get
occasion	cause
on a few occasions	occasionally
on behalf of	for
on the basis of	by, from, because
on the grounds that	because

Instead of:	Try:
on the occasion of	when
on the part of	for, among
on the subject of	about
optimum	the most for the least
orientated	oriented
outside of	outside
owing to the fact that	because
participate	take part
per diem, per annum	per day, a year
pertain	about, on
peruse	read
place emphasis on	emphasize
possess	have
preventative	preventive
prior to	before
proceed	go
procure	get
provided, providing	if
regarding	about
realize a savings of	save
reimburse	pay
related with, relates to, relating to, relative to	on, about
render aid or assistance	help
reported	said, told
resided	lived
residence	house, apartment
responded	answered, said
sibling	brother, sister
subject matter	subject, topic
submit	send, give
subsequently, subsequent to	later, afterward, then, next
sufficient	enough
summarization	summary
sustained	received
take action	act
terminate, terminated	end, ended, stopped, ending
the question as to whether	whether
the reason is due to	because
thereafter	after that, then
therein	there, in it
thereof	of it
thereupon	then
this is a subject that	this subject
transmit	send
transported	took, drove
under date of	on
under the circumstances	because
until such time as	until
utilization, utilize	use
vehicle	car, truck
visualize	see, think of, imagine
whereby	by which, so that
wherein	in which, where, when
whether or not	whether
wish to advise, wish to state	(avoid, do not use)
with regard to, with reference to, with respect to	about, regarding, concerning
with the result that	so that
without further delay	immediately, soon, quickly
would seem, would appear	seem/appear (try to avoid these)

ces are generally topic sentences in paragraphing. If you take good notes during the interview process, you should have no trouble understanding paragraphing.

A paragraph is a group of sentences that tells about one topic. The topic sentence tells what the sentences in the paragraph are about. Usually, it is the first sentence in the paragraph. Paragraphing is a method of alerting the reader to a shift in focus in the report.

Steps in Writing a Paragraph

First, your notes provide the key or topic sentence and the outline for the paragraphs. Check for completeness and rearrange sentences if necessary.

Second, write the paragraph in active-voice style, using 12- to 15-word sentences. Paragraphs in police reports generally have five to seven sentences or approximately 100 words. However, it is acceptable in police reports to write one- or two-sentence paragraphs. One- or two-sentence paragraphs are used to mark a transition in reports, from one topic or section to another: typically, going from the interview of the victim to the interview of a witness.

Indent the beginning of each paragraph or skip one or two lines between paragraphs.

Unity: Preserve the unity of the paragraph. A paragraph should develop a single topic, the key sentence. Every sentence in the paragraph should contribute to the development of that single idea.

Coherence: Compose the paragraph so it reads coherently. Coherence makes it easy for the reader to follow the facts and events. It reflects clear thinking by the report writer. A clearly stated chronological order of events makes the paragraph, and therefore the report, coherent.

Development: Paragraphs should be adequately developed. The first step is to consider the central idea. Present examples or specific quotes. Include relevant facts, details, or evidence. Explore and explain the causes of an event or the motives of the suspect. The result may be an explanation of how the event occurred. Finally, describe the scene, injuries, or other pertinent information.

Consider the following suggestions:

Repeat key sentences from paragraph to paragraph.

Use pronouns in place of key nouns.

Use "pointing words," for example, *this, that, these,* and *those.*

Use "thought-connecting words," such as *however, moreover, also, nevertheless, therefore, thus, subsequently, indeed, then,* and *accordingly.*

Arrange sentences in chronological order.

Third, proofread your work. If necessary, correct mistakes and rewrite the paragraph or report.

USE OF TAPE RECORDERS

You may consider using a tape recorder for both note taking and report dictation.

Note Taking

The use of a tape recorder for field note taking is generally discouraged. The biggest problem with tape recording field notes is that you capture too much unnecessary information. You may elect to use a tape recorder for note taking if you are interviewing a suspect in an involved or serious crime.

If you use a tape recorder for interviews, at the beginning of the interview always include the following:

Your name
Rank
Department
Date and time
Case number and type of case

Tape recorders may play an important role in obtaining unsolicited suspect statements where there is no violation of their constitutional rights. Under current case law, a suspect has no reasonable expectation to privacy in a police car. Therefore, officers may place a tape recorder in a patrol car and record suspect conversations.

Remember, there is no substitute for good note taking.

Report Dictation

You will find that dictating reports is much easier if you follow the three-step interview and scratch outline note-taking methods. The combination of these two methods is called a *dictation tree.* Some agencies now use dictation systems that have limited word-processing capabilities.

Eleven Dictation Tips

1. Organize your thoughts by reviewing your notes.
2. Relax for a few minutes after reviewing your notes.
3. When you first begin to dictate, state the type of report (what form to use) and your name and badge number.
4. Follow the order of the blocks on the form.
5. Slow your speech slightly.
6. Speak clearly; spell out names and words that are not easily understood.
7. Do not lose concentration: Don't try to listen for radio calls and the like.
8. Do not smoke, drink, chew gum, or eat during dictation.
9. If you make a mistake, pause; then tell the operator that you need to make a correction.
10. When finished, restate your name and badge number.
11. Use simple courtesy, "Thank you, operator."

CHAPTER REVIEW

In this chapter you learned to apply your skills and knowledge to write police reports. You can define and explain interpersonal communications, identify the five parts of the report-writing process, define chronological order, identify and write active voice sentences, identify appropriate word usage for police reports, identify the advantages of first versus third person, and properly use a tape recorder for note taking and report dictation.

DISCUSSION QUESTIONS

1. Explain why interpersonal communication skills are important to police officers.
2. Identify two of the four reasons we communicate, and discuss their importance and meaning.
3. Define and explain nonverbal communication.
4. What are the three steps in the three-step interview method and what does each accomplish?
5. Explain the value of three-step interviewing and scratch outlining when writing paragraphs.

EXERCISES

Revise the following sentences so they are clear, concise, and jargon and slang free.

1. The subject exited the stolen vehicle post hastily.

   _____

2. A theft in amount of $34.83 was reported.

   _____

3. The officer detected the odor of burning marijuana.

   _____

4. Officers contacted Lewis at his home.

   _____

5. The detective named Robinson as their primary suspect on account of the fact that his fingerprints matched those detectives found at the crime scene.

   _____

   _____

Replace each of the following words or phrases with a simple word or phrase:

Adjacent to   _____

Altercation   _____

Transported   _____

Observed   _____

Ascertain   _____

Choose the correct word in each of the following sentences:

1.   The (affect/effect) of the medication began to (affect/effect) his judgment.

2.   The burglar parked the car in the (alley/ally).

3.   The driver said he could not (brake/break) in time to avoid the accident.

4.   As the mortally wounded victim was (dying/dyeing), he named his assailant.

5.   The building was a strange (local/locale) for the gathering.

6.   Always use a (stationary/stationery) object when you need a point of reference.

7.   The investigators made a (through/thorough/through) investigation.

8. Officers will (advice/advise) the suspects of their rights.

9. The forgery suspect tried to (altar/alter) his appearance.

10. The whole neighborhood could (breath/breathe) easier once the police caught the escaped murderer.

Determine if these sentences are active or passive voice. If they are passive, rewrite them in active voice.

1. All the money was given to the suspect by the teller.

   _____

2. The suspect took the money and ran out the door.

   _____

3. A radio broadcast of the suspect's description was put out by me.

   _____

4. Two blocks away the suspect was stopped by Officer Wright.

   _____

5. The suspect was identified by the witness.

   _____

STUDENT WORKBOOK

Chapter One PARTS OF SPEECH

Identify the parts of speech in the following sentences:

1. The officers carefully patrolled the neighborhood after the shooting incident.
2. No one could easily identify the three suspects from the descriptions.
3. Officer Sanchez left the patrol car and searched the area on foot.
4. Mrs. Ngyun told the officers, "I would recognize the thief the minute I saw him again."
5. Mr. Irving said he was very cautious when he opened the door because he smelled smoke.
6. No patrol should be routine for any officer.
7. Witherspoon denied his guilt in the murder but admitted his hatred of the store owner.
8. The investigators will store the evidence for the upcoming trial in the evidence locker.
9. The crowd demanded justice, but really wanted revenge.
10. Two of the suspects surrendered inside the building, but the third suspect had quietly sneaked behind some trash cans in the alley before Officer Brooker caught him.

Chapter Two SENTENCE ELEMENTS

COMPLETE SENTENCES

Label the following sentences as complete (C), fragment (F), or run-on (R). Correct all fragments and run-ons.

_____ **1.** Halt!

_____

_____

_____ **2.** The two officers left at 0300, what time is it now?

_____

_____

_____ **3.** Until the investigators were able to sift through the evidence and learn the truth.

_____

_____

_____ **4.** Mario was reluctant to testify because the gang members frightened him; yet, he knew his story could save the defendant's life.

_____

_____

_____ **5.** Although Riley practiced shooting and carefully cleaned his gun after practice on the pistol range, and made every effort to improve.

_____

_____

_____ **6.** Why he did it he didn't know, he knew he couldn't get away with it.

_____

_____

_____ **7.** Although dazed from the injuries she received in the accident, Ms. Demont was still able to help the occupants from the other car.

_____

_____

_____ **8.** The investigator asked the bystanders if anyone had seen anything but no one wanted to get involved they were afraid of retaliation.

_____

_____

_____ 9. Train your mind to work efficiently and to catch minor mistakes.

_____

_____

_____ 10. Remember to check your reports for sentence fragments and run-ons
because an otherwise excellent police officer can appear less than
competent if his or her reports contain these common writing errors.

_____

_____

SUBJECT AND VERBS

Underline the simple subject once and the verb twice.

1. A picture of the suspect appeared in the paper.

2. The gang members scrubbed the wall, sanded it, and repainted it just to
    remove the graffiti.

3. There must have been fifteen witnesses to the bank robbery.

4. Write your notes clearly the first time.

5. Magazines, beer bottles, and partially smoked cigarettes were scattered
    around the room.

Circle the correct form of the verb in the following sentences:

1. Neither the suspect nor his accomplices (was/were) caught.
2. The mayor's use of statistics (make/makes) him sound important.
3. The van, but not its contents, (was/were) recovered.
4. The squad (has/have) chosen a new spokesperson.
5. Either the investigators or their captain (need/needs) to issue a statement.

Revise the following sentences:

1. There was only the suspect and the complainant in the room at the time.

    _____

    _____

2. Any one of those people were capable of committing the crime.

    _____

    _____

3. We need more traffic officers at Grand and Main; is any available?

    _____

    _____

4. Each of his fellow officers have contributed to the fund.

    _____

    _____

5. Among the suspects was a local pimp, a pusher, and an ex-con.

    _____

    _____

6. The report gave an account of the incident, and then the investigator asks more questions.

    _____

    _____

7. They commit the crime at 0440 hours and stole the car at 0500 hours.

    _____

    _____

8. The burglar quietly opens the door and then walked quickly down the hall.

    _____

    _____

9. The investigators kept trying to call the victim for weeks, but they didn't know he's already moved.

    _____

    _____

10. The hit-and-run driver went home and tells her husband what happened.

    _____

    _____

IRREGULAR VERBS

In the space provided, write the correct past tense form of the verb.

1. The witness said she (see) _____ the suspect enter the house.
2. He had (steal) _____ things in the past.

3. The investigators have (do) _____ all the investigations work and (write) _____ all the reports.

4. He should have (bring) _____ all the information with him.

5. He (drink) _____ so much beer that he had too much to drive home.

6. Rookies sometimes find they have not (take) _____ enough notes to write their narratives.

7. The captain had (give) _____ the new officers instructions and (show) _____ them what to do.

8. Officer Windom had already (eat) _____ when she received the call.

9. The burglar was (go) _____ before the sleepy residents (know) _____ that anything had been (take) _____ .

10. The body (lie) _____ on the floor but the coroner could not tell exactly how long it had (lie) _____ there.

DIRECT OBJECTS, INDIRECT OBJECTS, AND SUBJECT COMPLEMENTS

Underline the direct objects and/or subject complements in the following sentences. Circle any indirect objects you may find.

1. The forensic department studied the evidence carefully.

2. Witnesses showed the officers the evidence.

3. The report was the main focus of the attorney's objections.

4. Officer Kim read the suspect his rights.

5. Captain Foster was the only officer available.

6. The injured driver was lying beside the road.

7. The racketeer grew more powerful and greedy as his influence increased.

8. The hostages remained calm throughout the ordeal.

9. The city council gave the department a new contract with better benefits.

10. Which investigator did you want?

MODIFIERS

Underline the misplaced or dangling modifiers in the following sentences. Then correct the errors. In some cases you will have to rewrite the sentences.

1. Officer Nova found six marijuana cigarettes outside the car rolled with toilet paper.

    _____

    _____

2. The traffic controller watched the nine-car crash that happened on their closed-circuit TV monitor.

    _____

    _____

3. Officer Lemon killed the dog that attacked her with a single shot.

    _____

    _____

4. The supervisor told her he needed someone who could type badly.

    _____

    _____

5. After years of being lost in a back room filing cabinet, Stanley P. Duefuss found all the old case records.

    _____

    _____

6. I saw that the police had captured the murderer in the evening paper.

    _____

    _____

7. Officer Clark confiscated the switchblade from the suspect with a carved ivory handle.

    _____

    _____

8. Officers saw a suspicious van parked behind the building with two occupants in it.

    _____

    _____

9. Once coated with plastic, no one could alter the new identification cards.

    _____

    _____

10. The attorney only questioned these witnesses.

    _____

    _____

Chapter Three PRONOUNS

Select the correct pronoun form from the choices in parentheses.

1. The watch commander asked each officer to list (his/their) duty preferences.
2. Investigators found everyone home except (he/him) and his father.
3. The lieutenant left the final decision up to (we/us) officers.
4. When an officer is always late to briefing, (you/they/he) should expect a reprimand.
5. (Officer Ramirez and he/Officer Ramirez and him) followed the suspect's car.
6. The new captain called Officer Ngyun and (I/me) into his office.
7. Everyone was giving the officer (their/his or her) opinion at the same time.
8. Each division needed to submit (its/it's) budget requests.
9. All of the expended rounds of ammunition (was/were) confiscated.
10. The customers (who/which) were injured received medical treatment.

Circle and correct the agreement problems in these sentences.

1. The officer caught the suspect, but he slashed him on the arm.

    _____

    _____

2. The witness said she saw someone near the door, but they didn't come in.

    _____

    _____

3. Several people saw the suspects get out of the car. They went into the old building.

    _____

    _____

4. If one wants to succeed, you must work hard.

_____

_____

5. One can work happily if you like what you're doing.

_____

_____

6. Anyone who violates the law should be aware of the risk they are taking.

_____

_____

7. Officer Clay told the victim that he could call his doctor if he didn't feel well.

_____

_____

8. That dog's owner should be jailed. He howls all night.

_____

_____

9. Sgt. Jones told Sgt. Jimenez he didn't have his job anymore.

_____

_____

10. When the rioters left the buildings, the bystanders threw rocks at them.

_____

_____

Chapter Four MECHANICS

Correct all capitalization errors in the following sentences:

1. The Witness said, "officer, I saw the red ford hit the pedestrian and get on the freeway going South."
2. Thomas said That he had seen the Burglar leaving Sam's market at 6 P.M.
3. the canine unit carried a German Shepherd.
4. Officers in the south believed the woman they were looking for had left georgia and was traveling toward the Northern part of the country.
5. Smithers, Vice-President of Blue Dot inc., discovered the broken window when he arrived.

6. The grateful community awarded the Lieutenant the medal of valor after the riots last summer.
7. The juvenile said his Mom loaned him the mercedes but, "Boy, will dad be mad."
8. The bogus press newspaper quoted dr. Arnold, the Prime Suspect, as saying He got his idea from reading the Book <u>the bloody hatchet.</u>
9. The student told Officers the victim, ms. Rodriguez, taught biology II on tuesday and thursday, but he didn't know what other Science courses she taught.
10. To get to the Police Station, turn left off highway 62, go North on elm st. for two blocks, and then turn Right at the corner of elm and academy rd.

Insert periods, question marks, and/or exclamation marks in the following sentences.

1. The victim, Dr Ashcraft, reported the narcotics theft on Nov 10
2. Hurry The suspect just left
3. At the end of the interview, Sgt Bruckner asked if there was anything else she remembered
4. Is there anything else you can tell us about the accident
5. Go to the front door and wait for the lieutenant's instruction

Insert commas where necessary.

1. You should tell Sergeant Merrill not Captain Greene about this.
2. I spoke to the victim and she said she had just returned home when she heard the sound of breaking glass.
3. After your preliminary investigation you may have to talk to some witnesses again.
4. The accident occurred on May 10 1989 in Denver Colorado.
5. Jenkins's suicide note referred to his relatives his depression his lack of success and his feelings of inadequacy.
6. While trying to escape the suspect tripped and fell.
7. Sergeant Washington is the prisoner in custody?
8. Mr. Blaney a neighbor said "I know I've seen the car before but I just can't remember where."
9. The address the suspect gave 101 North Regency Avenue Apt. C Juneau Alaska didn't exist.
10. The property somehow was stored in the chief's office rather than in the property room.

Add semicolons and colons to the following:

1. He was supposedly extradited actually, he never left town.
2. The officer questioned three witnesses Lois Lane, the secretary Clark Kent, the reporter and Jimmy Olson, the copy boy.
3. Within three minutes of receiving the call, officers arrived however, the burglars were already gone.

4. After the alarm went off at 242 A.M., the thieves had only three minutes to finish the job and get out of the building.
5. Because the kidnappers were so inept, they wrote To Whom It May Concern on their ransom letter.

Add apostrophes and quotation marks to the following:

1. They found the rare, first edition copy of Mark Twains story The Mysterious Stranger after ten years worth of searching.
2. The defense attorney asked, Isnt it true, Dr. Simpson, that you prescribed the medication?
3. Dont waste a second. The sergeant will go bonkers if were late.
4. The thieves took some mens clothes, two childrens bikes, and a womans diamond ring.
5. Officer Jacksons reports are always well written and punctual, but why does she use so many *thens*?

Add hyphens and dashes to the following:

1. Woodall said his father in law caused the fight.
2. Deputies conducted a house to house search for the missing child.
3. About three fourths of the officers attended the meeting.
4. On pages 14 16 you will find the list of the twenty seven victims involved in the swindle.
5. He is the prime suspect the only suspect in the murder of the editor in chief of the newspaper.

Add parentheses, underlining, and periods (for ellipses) in the following:

1. Tierney told officers he paid eighty dollars $80 for the counterfeit copy of the painting A Bowl of Cherries.
2. The jury foreman read the verdict: "We the jury in the case of the State versus Anthony Adverse on the charge of fraud find the defendant guilty."
3. Sometimes it is difficult to tell his *i*'s from his *e*'s.
4. Officers gave the citizen three options: 1 settle the disagreement calmly, 2 call his lawyer, or 3 sign a formal complaint.
5. Dr. Cole Slab what an appropriate name for a medical doctor said the unidentified victim was DOA dead on arrival.

Select the correct usage of numbers in the following:

1. Narcotics officers confiscated approximately 15/fifteen pounds of cocaine.
2. They also logged into evidence $15,863/fifteen thousand eight hundred sixty-three dollars found at the scene.
3. 7/Seven suspects were arraigned on various charges.
4. The youngest suspect was 18/eighteen years old, and the oldest suspect was 62/sixty-two.

5. Investigators estimated the confiscated drugs represented only 25/twenty-five percent of the total amount shipped to the suspect.

Chapter Five SPELLING

Make the following words plural:

woman _____   crash _____

patch _____   building _____

thief _____   deer _____

tattoo _____   foot _____

fireman _____   city _____

Add the suffixes -ed and -ing to the following words:

confer _____   quarrel _____

occur _____   hop _____

burglarize _____   begin _____

step _____   copy _____

accelerate _____   testify _____

Add -able to the following:

agree _____   change _____

read _____   train _____

regret _____

Add -ness to the following:

close _____   sad _____

happy _____   same _____

sick _____

Circle the correctly spelled words from the following:

1. aggravated   agravated   aggrevated
2. argumentive   argumentative   arguementative
3. cematery   cemetary   cemetery
4. disturbance   disturbence   disterbance
5. lewtenant   lieutenent   lieutenant
6. payed   payd   paid
7. proceedure   procedur   procedure

8.	simular	similar	similiar
9.	unconscious	unconsious	unconscience
10.	writting	writeing	writing

Circle the correct homonym in the following:

1. The (*affect, effect*) of the medication began to (*affect, effect*) his judgment.

2. Murder is a (*capital, capitol*) offense.

3. Investigators were (*ceiling, sealing*) the building when the (*ceiling, sealing*) caved in.

4. They could not (*elicit, illicit*) any answer from the man about his (*elicit, illicit*) gambling habits.

5. The point of entry was the (*hole, whole*) in the roof.

6. He felt a sharp (*pain, pane*) in his hand after he broke the (*pain, pane*).

7. That recruit's (*principal, principle*) fault is not being able to think clearly under pressure.

8. Hammerfield said he was (*threw, through*) arguing and (*threw, through*) the bottle at Martinez.

9. He was (*to, too, two*) slow (*to, too, two*) load the last (*to, too, two*) rounds in the shotgun.

10. (*Your, You're*) letter of commendation proves (*your, you're*) an exemplary officer.

Chapter Six POLICE REPORTS

Circle the correct answer for each of the following:

1. The definition of a report is:
   A. Any documentation on a departmental form.
   B. Only documentation that is signed by a victim.
   C. Only documentation of crimes and not other events.

2. An arrest report must include:
   A. The chain of evidence.
   B. A request for forensic examination of evidence.
   C. Probable cause to stop, detain, and arrest the suspect.

3. Crime reports are completed:
   A. Only when you've recovered the stolen property.
   B. Your preliminary investigation concludes a crime has been committed.
   C. You need to document your daily activity.

4. Supplemental reports are completed:
   A. To record information you discover after the original report has been filed.
   B. To record the amount of time you've spent on the investigation.
   C. To record the number of similar crimes in the same area.

5. Police reports are written for the following reasons:
   A. To document criminal investigations.
   B. To provide reference material and historical data.
   C. Both A and B.
6. The source documents for crime analysis are:
   A. Crime reports.
   B. Memorandums.
   C. Daily activity reports.
7. The report-writing audience is made up of the following:
   A. Only police officers from your agency.
   B. Police officers, judges, administrators, and the media.
   C. Only those people directly at the scene of the crime.
8. The definition of accuracy in police reports is:
   A. Correct spelling and grammar.
   B. The use of correct report form.
   C. In exact conformity to fact: errorless.
9. The definition of clear in police reports means:
   A. Plain or evident to the mind of the reader.
   B. Understandable to anyone with an eighth grade education.
   C. Using terms common to law enforcement professionals.
10. Concise in a police report means:
    A. Short and to the point.
    B. Less than one page handwritten.
    C. Say as much as possible in as few words as possible.

Chapter Seven REPORT WRITING TECHNIQUES

DEADWOOD WORDS

Replace each of the following words or phrases with a simple word or phrase.

at this point in time   _____

contacted   _____

in the event of   _____

for the reason that   _____

during the course of   _____

a limited quantity of   _____

contingent upon receipt of   _____

render aid or assistance   _____

due to the fact that   _____

ameliorate   _____

WORD MEANINGS AND USAGE

Choose the correct word for the blanks in each sentence.

1. The officers told Higgins they would (cite/site/sight) him if he came on the construction (cite/site/sight) again.
2. The attorney wondered if the drunk would be a (credible/creditable) witness.
3. The motorist was stranded in the (desert/dessert).
4. Don't (loose/lose) sight of the suspects.
5. The (nosey/noisy) neighbor complained about the party.
6. The suspect became (quiet/quit/quite) when the witness identified her.
7. Stolen cars are often (stripped/striped) before anyone finds them.
8. Rodriguez (waived/waved) his rights and confessed.
9. We (were/where) working in an area (wear/where) we had to (where/wear) special uniforms.
10. The officers had (all ready/already) arrived before the burglars were (all ready/already) to leave.

Revise the following sentences so they are clear, concise, and jargon and slang free.

1. The witness indicated to the investigating officer that she had indeed observed the suspected culprit depart from the jewelry store just prior to the time that the officers responded to the location.

    _____

    _____

2. The assault victim was transported to the hospital where slides were taken of his head which was booked as evidence at the station.

    _____

    _____

3. Officer Kim searched the trunk with negative results.

    _____

    _____

4. The victim, Sally Shields, was really shook, but she was still cool about the details of what had gone down.

    _____

    _____

5. The neighbors were of the opinion that the investigators had left no stone unturned in the relentless search for the fugitive from justice.

    _____

    _____

6. Upon her arrival, Officer Beckworth ascertained that there were seven witnesses she had to contact in order to commence her investigation into the physical altercation between Samuels and Whitson.

_____

_____

7. Bound, gagged, and trussed up in a denim bag, with plugs in her ears and tape over her eyes, the victim Miss Sarah Sharp told yesterday how she was kidnapped.

_____

_____

8. She was brought down to the station by Detective Jorgenson for the purpose of the identification of the suspect.

_____

_____

9. The gang was made up of six Gray Devils from Frisco, an equal number of representatives from Sacramento, and the same amount of members from L.A.

_____

_____

10. Olson was verbally advised by this officer to give this officer the baton belonging to said officer.

_____

_____

Change the following passive voice sentences to active voice sentences.

1. Latent fingerprints were found on the empty bottle.

_____

_____

2. The monthly crime report was completed by Chief Bowman and submitted to the City Council.

_____

_____

3. Officer Charles was informed by Sgt. Lasiter that the search of the scene had been conducted by the laboratory technicians.

_____

_____

4. The juvenile suspect was found crouching behind some garbage cans in the alley by Officer Owens.

   _____

   _____

5. It was determined by the officers that entrance was gained by the burglars through a side door.

   _____

   _____

List of Irregular Verbs

Present	Past	Past Participle
be	was/were	been
bear	bore	borne
beat	beat	beaten
become	became	become
begin	began	begun
bend	bent	bent
bet	bet	bet
bid (offer)	bid	bid
bid (command)	bade	bidden
bite	bit	bitten
bleed	bled	bled
blow	blew	blown
break	broke	broken
bring	brought	brought
build	built	built
burst	burst	burst
buy	bought	bought
can	could	no past participle
cast	cast	cast
catch	caught	caught
choose	chose	chosen
cling	clung	clung
come	came	come
cost	cost	cost
creep	crept	crept
cut	cut	cut
deal	dealt	dealt
dive	dove/dived	dived
do (does)	did	done
draw	drew	drawn
drink	drank	drunk
drive	drove	driven
eat	ate	eaten
fall	fell	fallen
feed	fed	fed
feel	felt	felt
fight	fought	fought
find	found	found
flee	fled	fled
fling	flung	flung
fly	flew	flown
forget	forgot	forgot (-ten)
forgive	forgave	forgiven
freeze	froze	frozen
get	got	got/gotten
give	gave	given
go	went	gone
grind	ground	ground
grow	grew	grown
hang	hung	hung
have	had	had
hear	heard	heard
hide	hid	hidden
hit	hit	hit
hold	held	held
hurt	hurt	hurt
keep	kept	kept
know	knew	known
lay	laid	laid
lead	led	led
leave	left	left
lend	lent	lent

Present	Past	Past Participle
let	let	let
lie	lay	lain
lose	lost	lost
make	made	made
may	might	no past participle
mean	meant	meant
meet	met	met
pay	paid	paid
put	put	put
quit	quit	quit
read	read	read
rid	rid	rid
ride	rode	ridden
ring	rang	rung
rise	rose	risen
run	ran	run
say	said	said
see	saw	seen
seek	sought	sought
sell	sold	sold
send	sent	sent
set	set	set
shake	shook	shaken
shed	shed	shed
shine	shone	shone
shoot	shot	shot
shrink	shrank	shrunk
shut	shut	shut
sing	sang	sung
sink	sank/sunk	sunk
sit	sat	sat
sleep	slept	slept
slide	slid	slid
slit	slit	slit
speak	spoke	spoken
speed	sped	sped
spend	spent	spent
spin	spun	spun
split	split	split
spread	spread	spread
spring	sprang/sprung	sprung
stand	stood	stood
steal	stole	stolen
sting	stung	stung
strike	struck	struck
swear	swore	sworn
sweep	swept	swept
swim	swam	swum
swing	swung	swung
take	took	taken
teach	taught	taught
tear	tore	torn
tell	told	told
think	thought	thought
throw	threw	thrown
thrust	thrust	thrust
wear	wore	worn
weave	wove	woven
win	won	won
wind	wound	wound
write	wrote	wrote

Note: Of is never a helping verb. If you have ever said or written,
"I could *of* done it," you meant, "I could *have* done it."

List of Nonaction Linking Verbs

is	may	has
are	can	had
was	must	will
were	could	shall
am	would	might
be	should	look
been	do	feel
being	does	seem
become	did	taste
became	have	appear

Practice Scenarios and Sample Reports

Directions: Write an appropriate narrative for each of the following scenarios. If you are using police department crime reports, complete all necessary portions of the form. If necessary, use today's date and time. You are the officer assigned to take the report.

Scenario 1   Bicycle Theft Report

You are sent to 14621 Spring Street, Riverdale, to take a bicycle theft report. Your reporting party/victim is:

       Victim: Sheila Marie Garvin DOB: 1-18-60
       Phones: (h) 891-1722   (w) 835-3802
Work Address: 212 North Main St., Corona, CA 92000
   Occupation: Office Manager Female-Caucasian

You interview Ms. Garvin and she tells you the following: She took her 10-year-old son to school at 7:45 A.M. She remembers seeing the bicycle in the driveway because her son had to move it out of the way. He put the bicycle on the front porch. After she took him to school, she stopped at the grocery store to buy some milk. She returned home at approximately 8:30 A.M.

When she got home, the bicycle was gone. She looked in the rear yard and garage, but couldn't find it. No one has permission to use the bicycle.

She described the bicycle as a boy's 26" Huffy All American, painted red and white with a black seat. She paid $185.00 for it.

Scenario 2   Minor in Possession of Alcohol

You park your police car and walk through Mt. Rushmore City Park, located at the corner of Park Drive and Mountain Avenue. It is a violation of city ordinance No. 3-18 to possess alcoholic beverages in a city park, and it is also a violation of state law for a minor (someone between the ages of 18-21 years) to possess an alcoholic beverage.

You walk toward the outdoor handball courts and see a young male subject sitting on the ground. He's got a can of Budweiser beer in his right hand. When you walk up to him, he tries to hide it behind his right leg.

At your request, the subject gives you his driver's license and hands you the can of beer. He admits it is his. His driver's license shows that he is only 19 years old.

You dump out the liquid in the can to make sure it looks and smells like beer. You arrest him for possession of an alcoholic beverage by a minor and release him on a citation at the scene.

> Suspect: James Michael Senecal DOB: 12-5-____
> Phones: (h) 981-7122
> Home Address: 1012 Marigold, Middletown, USA
> Occupation: unemployed Male-Caucasian

Scenario 3 Residential Burglary

You are sent to 24906 Mayberry Road, Hometown, to take a residential burglary report. Your reporting party/victim is:

> Victim: Allen Mark Talbot DOB: 2-23-39
> Phones: (h) 918-1227 (w) 235-0242
> Work Address: 1039 Broadway, Hometown, CA 90000
> Occupation: Engineer Male-Caucasian

Talbot locked all the windows and doors to his house and left for work at 7:45 this morning. He came home at about 5:30 P.M. and found the front door open. When he went inside, he discovered someone had torn his house apart. The furniture was overturned, drawers were emptied on the floor, and clothes removed from the closet.

He checked the dresser in the master bedroom and found approximately three hundred dollars in cash missing. He also found his Rolex watch was missing from the night table by his bed.

You check the residence for physical evidence and to find the point of entry. You find the sliding glass door to the master bedroom has been pried open and there's a small screwdriver left on the patio. You check the neighborhood for witnesses and can't find anyone who saw the suspect.

Mr. Talbot lives alone, his wife died two years ago, and doesn't have any idea who broke into his house.

SAMPLE NARRATIVES

1 Petty Theft

Victim Garcia left for work this morning at 7:30 A.M. He remembered seeing his potted plant on the front porch next to the door. When he came home for lunch at 12:30 P.M. the plant was gone. He didn't have any idea who took the plant.

2 Assault and Battery

Today at 4:50 P.M. Deputy Arzate and I were dispatched to a report of two men fighting at the Sip-n-Bull Bar. When we arrived, Laughlin and Lindstrom were fighting outside the bar. Arzate and I separated the two men and I interviewed Laughlin first. Deputy Arzate stayed with Lindstrom.

Laughlin told me the following: He and Lindstrom met in the bar about six months ago. They usually meet there about twice a week, especially during the football season. Laughlin said they make "friendly" wagers on the upcoming games. In the past two weeks, Laughlin has won $250 from Lindstrom, but Lindstrom hasn't paid his debt.

Today they started drinking beer about 2:00 P.M. and talking about this week's games. When Laughlin asked Lindstrom to pay him, they began to argue. Finally, Lindstrom pushed Laughlin off the bar stool. When Laughlin got up, Lindstrom used his right hand in a clenched fist to punch Laughlin in the face. The punch knocked Laughlin to the floor.

Laughlin said he was afraid of Lindstrom. He got up and ran out of the bar, but Lindstrom caught him outside. Laughlin said the police arrived at about the same time.

I explained private person arrest to Laughlin and he said he was willing to arrest Lindstrom.

I walked up to Lindstrom and said, "How's it going?" Lindstrom spontaneously said, "I'm sorry. I shouldn't have punched his lights out. I know I owe him the money, but I've been out of work for two months." When I told Lindstrom that Laughlin was going to arrest him, Lindstrom said, "I won't cause you any trouble, I've been arrested before."

Laughlin arrested Lindstrom for assault and battery. I handcuffed Lindstrom and booked him at the county jail.

3 Residential Burglary

I interviewed victim Brower and she told me the following: At 7:30 A.M. she left her house and was sure all windows and doors were locked. She returned at 9:30 A.M. and unlocked the front door to let herself into the house.

She saw the TV and stereo were missing from the living room and nothing else in the house was disturbed. She went into the kitchen and saw the rear door was open. She didn't see any signs of forced entry or damage to the door. Ms. Brower said she didn't have any idea who broke into her house.

I called for an identification technician to check for latent fingerprints and other physical evidence.

When I left the residence, a neighbor and witness, Mrs. White, came up and told me what she saw. Her house is across the street and from her kitchen window she can see the front of Brower's residence.

White said she saw two men sitting in a small pickup truck in front of Brower's house at about 8:30 A.M. The two men talked for awhile; then the older one got out of the driver's door and went around to the back of Brower's house. In a few minutes he came out the front door carrying the television. He put the TV in the back of the truck. The second man got out of the truck and went through the front door into the house and came out with the stereo. He got in the cab of the truck with the stereo. The older man was already in the driver's seat. The men drove north toward Vineyard Road.

White said she didn't recognize the men or the truck, but could identify both men and the truck if she saw them again.

ANSWER KEY FOR STUDENT WORKBOOK

Chapter One PARTS OF SPEECH

Identify the parts of speech in the following sentences:

1. The officers carefully patrolled the neighborhood after the shooting incident.
   art. noun adverb   verb   art. noun   prep. art. adj.   noun

2. No one could easily identify the three suspects from the descriptions.
   pronoun verb adv.   verb   art. adj. noun prep. art.   noun

3. Officer Sanchez left the patrol car and searched the area on foot.
   noun   verb art. adj. noun conj. verb   art. noun prp. noun

4. Mrs. Ngyun told the officers, "I would recognize the thief the minute I
   noun   verb art. noun   pron. verb verb   art. noun art. noun   pron.
   saw him again."
   verb pron. adv.

5. Mr. Irving said he was very cautious when he opened the door because he
   noun   verb prn. verb adv. adj.   conj. prn. verb art. noun conj.   prn.*
   smelled smoke.
   verb   noun

6. No patrol should be routine for any officer.
   adj. noun verb verb adj. prep. adj. noun

7. Witherspoon denied his guilt in the murder but admitted his hatred of the
   noun   verb prn. noun prep. art. noun conj. verb   prn. †noun prep. art.
   store owner.
   adj.   noun

8. The investigators will store the evidence for the upcoming trial in
   art. noun   verb verb art. noun   prep. art. adj.   noun prep.

the evidence locker.
art. adj.   noun

9. The crowd demanded justice, but really wanted revenge.
   art. noun   verb   noun conj. adv. verb   noun

10. Two of the suspects surrendered inside the building, but the third suspect
    noun prep. art. noun   verb   prep. art. noun   conj. art. adj. noun
    had quietly sneaked behind some trash cans in the alley before
    verb adv.   verb   prep.   adj. adj. noun prep. art. noun prep.
    Officer Brooker caught him.
    noun   verb prn.

*In this case, *when* joins two clauses.
†*this* is a possessive pronoun used as an adjective.

Chapter Two SENTENCE ELEMENTS

COMPLETE SENTENCES

Label the following sentences as complete (C), fragment (F), or run-on (R). Correct all fragments and run-ons.
Answers to rewritten sentences may vary.

<u> C </u>   1. Halt!

<u> R </u>   2. The two officers left at 0300, what time is it now?
                   <u>The two officers left at 0300. What time is it now?</u>

<u> F </u>   3. Until the investigators were able to sift through the evidence and learn the truth.
                   <u>The investigators couldn't make an arrest until they sifted through the evidence and learned the truth.</u>

<u> C </u>   4. Mario was reluctant to testify because the gang members frightened him; yet, he knew his story could save the defendant's life.

<u> F </u>   5. Although Riley practiced shooting and carefully cleaned his gun after practice on the pistol range, and made every effort to improve.
                   <u>Although Riley practiced shooting, carefully cleaned his gun after practice on the pistol range, and made every effort to improve, he was unable to qualify.</u>

<u> R </u>   6. Why he did it he didn't know, he knew he couldn't get away with it.
                   <u>Why he did it he didn't know. *He* knew he couldn't get away with it.</u>

<u> C </u>   7. Although dazed from the injuries she received in the accident, Ms. Demont was still able to help the occupants from the other car.

<u> R </u>   8. The investigator asked the bystanders if anyone had seen anything but no one wanted to get involved they were afraid of retaliation.
                   <u>The investigator asked the bystanders if anyone had seen anything, but no one wanted to get involved. *They* were afraid of retaliation.</u>

<u> C </u>   9. Train your mind to work efficiently and to catch minor mistakes.

  C   **10.** Remember to check your reports for sentence fragments and run-ons because an otherwise excellent police officer can appear less than competent if his or her reports contain these common writing errors.

SUBJECT AND VERBS

Underline the simple subject once and the verb twice.

1. A <u>picture</u> of the suspect <u><u>appeared</u></u> in the paper.
2. The gang <u>members</u> <u><u>scrubbed</u></u> the wall, <u><u>sanded</u></u> it, and <u><u>repainted</u></u> it just to remove the graffiti.
3. There <u><u>must have been</u></u> fifteen <u>witnesses</u> to the bank robbery.
4. <u>(You)</u> <u><u>Write</u></u> your notes clearly the first time.
5. <u>Magazines</u>, beer <u>bottles</u>, and partially smoked <u>cigarettes</u> <u><u>were scattered</u></u> around the room.

Circle the correct form of the verb in the following sentences:

1. Neither the suspect nor his accomplices (was/*were*) caught.
2. The mayor's use of statistics (make/*makes*) him sound important.
3. The van, but not its contents, (*was*/were) recovered.
4. The squad (*has*/have) chosen a new spokesperson.
5. Either the investigators or their captain (need/*needs*) to issue a statement.

Revise the following sentences:

1. There was (*were*) only the suspect and the complainant in the room at the time.
2. Any one of those people were (*was*) capable of committing the crime.
3. We need more traffic officers at Grand and Main; is (*are*) any available?
4. Each of his fellow officers have (*has*) contributed to the fund.
5. Among the suspects was (*were*) a local pimp, a pusher, and an ex-con.
6. The report gave an account of the incident, and then the investigator asks (*asked*) more questions.
7. They commit (*committed*) the crime at 0440 hours and stole the car at 0500 hours.
8. The burglar quietly opens (*opened*) the door and then walked quickly down the hall.
9. The investigators kept trying to call the victim for weeks, but they didn't know he's (*he'd or he had*) already moved.
10. The hit-and-run driver went home and tells (*told*) her husband what happened.

IRREGULAR VERBS

In the space provided, write the correct past tense form of the verb:

1. The witness said she (see) <u>saw</u> the suspect enter the house.
2. He had (steal) <u>stolen</u> things in the past.

3. The investigators have (do) <u>done</u> all the investigations work and (write) <u>written</u> all the reports.
4. He should have (bring) <u>brought</u> all the information with him.
5. He (drink) <u>drank</u> so much beer that he had too much to drive home.
6. Rookies sometimes find they have not (take) <u>taken</u> enough notes to write their narratives.
7. The captain had (give) <u>given</u> the new officers instructions and (show) <u>shown</u> them what to do.
8. Officer Windom had already (eat) <u>eaten</u> when she received the call.
9. The burglar was (go) <u>gone</u> before the sleepy residents (know) <u>knew</u> that anything had been (take) <u>taken</u>.
10. The body (lie) <u>lay</u> on the floor but the coroner could not tell exactly how long it had (lie) <u>lain</u> there.

DIRECT OBJECTS, INDIRECT OBJECTS, AND SUBJECT COMPLEMENTS

Underline the direct objects and/or subject complements in the following sentences. Circle any indirect objects you may find.

1. The forensic department studied the <u>evidence</u> carefully.
2. Witnesses showed the *officers* the <u>evidence</u>.
3. The report was the main <u>focus</u> of the attorney's objections.
4. Officer Kim read the *suspect* his <u>rights</u>.
5. Captain Foster was the only <u>officer</u> available.
6. The injured driver was lying beside the road. *none*
7. The racketeer grew more <u>powerful</u> and <u>greedy</u> as his influence increased.
8. The hostages remained <u>calm</u> throughout the ordeal.
9. The city council gave the *department* a new <u>contract</u> with better benefits.
10. Which <u>investigator</u> did you want?

MODIFIERS

Underline the misplaced dangling modifiers in the following sentences. Then correct the errors. In some cases you will have to rewrite the sentences.

1. Officer Nova found six marijuana cigarettes outside the car <u>rolled with toilet paper</u>.
2. The traffic controller watched the nine-car crash <u>that happened on their closed-circuit TV monitor</u>.
3. Officer Lemon killed the dog that attacked her <u>with a single shot</u>.
4. The supervisor told her he needed someone <u>who could type badly</u>.
5. <u>After years of being lost in a back room filing cabinet</u>, Stanley P. Duefuss found all the old case records.
6. I saw that the police had captured the murderer <u>in the evening paper</u>.
7. Officer Clark confiscated the switchblade from the suspect <u>with a carved ivory handle</u>.

8. Officers saw a suspicious van parked behind the building <u>with two occupants in it</u>.

9. <u>Once coated with plastic</u>, no one could alter the new identification cards.

10. The attorney <u>only</u> questioned these witnesses.

Chapter Three PRONOUNS

Select the correct pronoun form from the choices in parentheses.

1. The watch commander asked each officer to list (*his*/their) duty preferences.

2. Investigators found everyone home except (he/*him*) and his father.

3. The lieutenant left the final decision up to (we/*us*) officers.

4. When an officer is always late to briefing, (you/they/*he*) should expect a reprimand.

5. (*Officer Ramirez and he*/Officer Ramirez and him) followed the suspect's car.

6. The new captain called Officer Ngyun and (I/*me*) into his office.

7. Everyone was giving the officer (their/*his or her*) opinion at the same time.

8. Each division needed to submit (*its*/it's) budget requests.

9. All of the expended rounds of ammunition (was/*were*) confiscated.

10. The customers (*who*/which) were injured received medical treatment.

Circle and correct the agreement problems in these sentences—*answers may vary*.

1. The officer caught the suspect, but *he* slashed *him* on the arm. <u>The officer caught the suspect, but the suspect slashed the officer's arm.</u>

2. The witness said she saw someone near the door, but *they* didn't come in. <u>The witness said she saw someone near the door, but he didn't come in.</u>

3. Several people saw the suspects get out of the car. *They* went into the old building. <u>Several people saw the suspects get out of the car. The suspects (*or* the people) went into the old building.</u>

4. If one wants to succeed, *you* must work hard. <u>If one wants to succeed, one must work hard.</u>

5. One can work happily if *you* like what *you're* doing. <u>One can work happily if one likes what one is doing.</u>

6. Anyone who violates the law should be aware of the risk *they are* taking. <u>Anyone who violates the law should be aware of the risk he or she is taking.</u>

7. Officer Clay told the victim that *he* could call *his* doctor. <u>Officer Clay told the victim that the victim could call a doctor.</u>

8. That dog's owner should be jailed. *He* howls all night. <u>That dog's owner should be jailed. The dog howls all night.</u>

9. Sgt. Jones told Sgt. Jimenez *he* didn't have *his* job anymore. <u>Sgt. Jones told Sgt. Jimenez Jones (or Jimenez) didn't have his job anymore.</u>

10. When the rioters left the buildings, the bystanders threw rocks at them. <u>When the rioters left the buildings, the bystanders threw rocks at the rioters.</u>

Chapter Four MECHANICS

Correct all capitalization errors in the following sentences:

1. The *w*itness said, "*O*fficer, I saw the red Ford hit the pedestrian and get on the freeway going *s*outh."
2. Thomas said *t*hat he had seen the *b*urglar leaving Sam's *M*arket at 6 P.M.
3. *T*he canine unit carried a German *s*hepherd.
4. Officers in the *S*outh believed the woman they were looking for had left Georgia and was traveling toward the *n*orthern part of the country.
5. Smithers, *v*ice-*p*resident of Blue Dot Inc., discovered the broken window when he arrived.
6. The grateful community awarded the *l*ieutenant the Medal of Valor after the riots last summer.
7. The juvenile said his *m*om loaned him the Mercedes but, "Boy, will *D*ad be mad."
8. The *B*ogus *P*ress newspaper quoted *D*r. Arnold, the *p*rime *s*uspect, as saying *h*e got his idea from reading the *b*ook <u>The Bloody Hatchet.</u>
9. The student told *o*fficers the victim, *M*s. Rodriguez, taught *B*iology II on Tuesday and *T*hursday, but he didn't know what other *s*cience courses she taught.
10. To get to the *p*olice *s*tation, turn left off *H*ighway 62, go *n*orth on *E*lm *S*t. for two blocks, and then turn *r*ight at the corner of *E*lm and Academy *R*d.

Insert periods, question marks, and/or exclamation marks in the following sentences.

1. The victim, Dr. Ashcraft, reported the narcotics theft on Nov. 10.
2. Hurry*!* The suspect just left.
3. At the end of the interview, Sgt. Bruckner asked if there was anything else she remembered.
4. Is there anything else you can tell us about the accident?
5. Go to the front door and wait for the lieutenant's instruction.

Insert commas where necessary.

1. You should tell Sergeant Merrill, not Captain Greene, about this.
2. I spoke to the victim, and she said she had just returned home when she heard the sound of breaking glass.
3. After your preliminary investigation, you may have to talk to some witnesses again.
4. The accident occurred on May 10, 1989, in Denver, Colorado.
5. Jenkin's suicide note referred to his relatives, his depression, his lack of success, and his feelings of inadequacy.
6. While trying to escape, the suspect tripped and fell.
7. Sergeant Washington, is the prisoner in custody?
8. Mr. Blaney, a neighbor, said, "I know I've seen the car before, but I just can't remember where."

9. The address the suspect gave, 101 North Regency Avenue, Apt. C, Juneau, Alaska, didn't exist.
10. The property somehow was stored in the chief's office rather than in the property room. *no additional commas*

Add semicolons and colons to the following:

1. He was supposedly extradited; actually, he never left town.
2. The officer questioned three witnesses: Lois Lane, the secretary; Clark Kent, the reporter; and Jimmy Olson, the copy boy.
3. Within three minutes of receiving the call, officers arrived; however, the burglars were already gone.
4. After the alarm went off at 2:42 A.M., the thieves had only three minutes to finish the job and get out of the building.
5. Because the kidnappers were so inept, they wrote To Whom It May Concern: on their ransom letter.

Add apostrophes and quotation marks to the following:

1. They found the rare, first edition copy of Mark Twain's story "The Mysterious Stranger" after ten years' worth of searching.
2. The defense attorney asked, "Isn't it true, Dr. Simpson, that you prescribed the medication?"
3. Don't waste a second. The sergeant will "go bonkers" if we're late.
4. The thieves took some men's clothes, two children's bikes, and a woman's diamond ring.
5. Officer Jackson's reports are always well written and punctual, but why does she use so many *then's*?

Add hyphens and dashes to the following:

1. Woodall said his father-in-law caused the fight.
2. Deputies conducted a house-to-house search for the missing child.
3. About three-fourths of the officers attended the meeting.
4. On pages 14-16 you will find the list of the twenty-seven victims involved in the swindle.
5. He is the prime suspect - the only suspect in the murder of the editor-in-chief of the newspaper.

Add parentheses, underlining, and periods (for ellipses) in the following:

1. Tierney told officers he paid eighty dollars ($80) for the counterfeit copy of the painting A Bowl of Cherries.
2. The jury foreman read the verdict: "We the jury in the case of the State versus Anthony Adverse on the changes of ... find the defendant guilty."
3. Sometimes it is difficult to tell his *i*'s from his *e*'s.
4. Officers gave the citizen three options: (1) settle the disagreement calmly, (2) call his lawyer, or (3) sign a formal complaint.

5. Dr. Cole Slab (what an appropriate name for a medical doctor) said the unidentified victim was DOA (dead on arrival).

Select the correct usage of numbers in the following:

1. Narcotics officers confiscated approximately *15*/fifteen pounds of cocaine.
2. They also logged into evidence *$15,863*/fifteen thousand eight hundred sixty-three dollars found at the scene.
3. *7*/*Seven* suspects were arraigned on various charges.
4. The youngest suspect was 18/*eighteen* years old, and the oldest suspect was 62/*sixty-two*.
5. Investigators estimated the confiscated drugs represented only *25*/twenty-five percent of the total amount shipped to the suspect.

Chapter Five SPELLING

Make the following words plural:

woman	women	crash	crashes
patch	patches	building	buildings
thief	thieves	deer	deer
tattoo	tattoos	foot	feet
fireman	firemen	city	cities

Add the suffixes *-ed* and *-ing* to the following words:

confer	conferred, conferring
occur	occurred, occurring
burglarize	burglarized, burglarizing
step	stepped, stepping
accelerate	accelerated, accelerating
quarrel	quarreled, quarreling
hop	hopped, hopping
begin	no *-ed*, beginning
copy	copies, copying
testify	testified, testifying

Add *-able* to the following:

agree	agreeable
read	readable
regret	regrettable
change	changeable
train	trainable

Add *-ness* to the following:

| close | closeness |
| happy | happiness |

sick	sickness
sad	sadness
same	sameness

Circle the correctly spelled words from the following:

1. *aggravated*   agravated   aggreved
2. argumentive   *argumentative*   arguementative
3. cematery   cemetary   *cemetery*
4. *disturbance*   disturbence   disterbance
5. lewtenant   lieutenent   *lieutenant*
6. payed   payd   *paid*
7. proceedure   procedur   *procedure*
8. simular   *similar*   similiar
9. *unconscious*   unconsious   unconscience
10. writting   writeing   *writing*

Circle the correct homonym in the following:

1. The (affect, *effect*) of the medication began to (*affect*, effect) his judgment.
2. Murder is a (*capital*, capitol) offense.
3. Investigators were (ceiling, *sealing*) the building when the (*ceiling*, sealing) caved in.
4. They could not (*elicit*, illicit) any answer from the man about his (elicit, *illicit*) gambling habits.
5. The point of entry was the (*hole*, whole) in the roof.
6. He felt a sharp (*pain*, pane) in his hand after he broke the (pain, *pane*).
7. That recruit's (*principal*, principle) fault is not being able to think clearly under pressure.
8. Hammerfield said he was (threw, *through*) arguing and (*threw*, through) the bottle at Martinez.
9. He was (to, *too*, two) slow (*to*, too, two) load the last (to, too, *two*) rounds in the shotgun.
10. (*Your*, You're) letter of commendation proves (your, *you're*) an exemplary officer.

Chapter Six POLICE REPORTS

Circle the correct answer for each of the following:

1. The definition of a report is:
   A. any documentation on a departmental form.
   B. only documentation that is signed by a victim.
   C. only documentation of crimes and not other events.
2. An arrest report must include:
   A. the chain of evidence.
   B. a request for forensic examination of evidence.
   C. probable cause to stop, detain, and arrest the suspect.

3. Crime reports are completed:
   A. only when you've recovered the stolen property.
   B. your preliminary investigation concludes a crime has been committed.
   C. you need to document your daily activity.

4. Supplemental reports are completed:
   A. to record information you discover after the original report has been filed.
   B. to record the amount of time you've spent on the investigation.
   C. to record the number of similar crimes in the same area.

5. Police reports are written for the following reasons:
   A. to document criminal investigations.
   B. to provide reference material and historical data.
   C. both A and B.

6. The source document for crime analysis is:
   A. crime reports.
   B. memorandums.
   C. daily activity reports.

7. The report-writing audience is made up of the following:
   A. only police officers from your agency.
   B. police officers, judges, administrators, and the media.
   C. only those people directly at the scene of the crime.

8. The definition of accuracy in police reports is:
   A. correct spelling and grammar.
   B. the use of the correct report form.
   C. in exact conformity to fact: errorless.

9. The definition of clear in police reports means:
   A. plain to evident to the mind of the reader.
   B. understandable to anyone with an eighth grade education.
   C. using terms common to law enforcement professionals.

10. Concise in a police report means:
    A. short and to the point.
    B. less than one page handwritten.
    C. say as much as possible in as few words as possible.

Chapter Seven REPORT WRITING TECHNIQUES

DEADWOOD WORDS

Replace each of the following words or phrases with a simple word or phrase.

at this point in time	now
contacted	called
in the event of	if
for the reason that	because
during the course of	during
a limited quantity of	few
contingent upon receipt of	when we receive

render aid or assistance	help
due to the fact that	because
ameliorate	improve

WORD MEANINGS AND USAGE

Choose the correct word for the blanks in each sentence.

1. The officers told Higgins they would (*cite*/site/sight) him if he came on the construction (cite/*site*/sight) again.
2. The attorney wondered if the drunk would be a (*credible*/creditable) witness.
3. The motorist was stranded in the (*desert*/dessert).
4. Don't (loose/*lose*) sight of the suspects.
5. The (*nosey*/noisy) neighbor complained about the party.
6. The suspect became (*quiet*/quit/quite) when the witness identified her.
7. Stolen cars are often (*stripped*/striped) before anyone finds them.
8. Rodriguez (*waived*/waved) his rights and confessed.
9. We (*were*/where) working in an area (wear/*where*) we had to (where/*wear*) special uniforms.
10. The officers had (all ready/*already*) arrived before the burglars were (*all ready*/already) to leave.

Revise the following sentences so they are clear, concise, and jargon and slang free— *answers may vary.*

1. The witness indicated to the investigating officer that she had indeed observed the suspected culprit depart from the jewelry store just prior to the time that the officers responded to the location.
   The witness told the officer she saw the suspect leave the jewelry store just before the officer arrived.
2. The assault victim was transported to the hospital where slides were taken of his head which was booked as evidence at the station.
   Medix ambulance took the victim to the hospital. Dr. Jones took x-rays of the victim's head. Officer Miller booked the x-rays as evidence.
3. Officer Kim searched the trunk with negative results.
   Officer Kim searched the trunk and didn't find anything.
4. The victim, Sally Shields, was really shook, but she was still cool about the details of what had gone down.
   The victim, Sally Shields, was upset; however, she was able to recall details of the event.
5. The neighbors were of the opinion that the investigators had left no stone unturned in the relentless search for the fugitive from justice.
   The neighbors thought the investigators completed a thorough search for the fugitive.
6. Upon her arrival, Officer Beckworth ascertained that there were seven witnesses she had to contact in order to commence her investigation into the physical altercation between Samuels and Whitson.

When Officer Beckwith arrived, she found seven witnesses to the fight between Samuels and Whitson.

7. Bound, gagged, and trussed up in a denim bag, with plugs in her ears and tape over her eyes, the victim Miss Sarah Sharp told yesterday how she was kidnapped.

Kidnap victim Sarah Sharp was bound, gagged, and tied in a denim bag. Her ears were plugged and tape placed over her eyes.

8. She was brought down to the station by Detective Jorgenson for the purpose of the identification of the suspect.

Detective Jorgenson brought her to the station to identify the suspect.

9. The gang was made up of six Gray Devils from Frisco, an equal number of representatives from Sacramento, and the same amount of members from L.A.

There were six Gray Devil gang members from San Francisco, six from Los Angeles, and six from Sacramento.

10. Olson was verbally advised by this officer to give this officer the baton belonging to said officer.

I told Olson to give me my baton.

Change the following passive voice sentences to active voice sentences—*answers will vary.*

1. Latent fingerprints were found on the empty bottle.

I found latent fingerprints on the bottle.

2. The monthly crime report was completed by Chief Bowman and submitted to the City Council.

Chief Bowman completed the monthly crime report and submitted it to the City Council.

3. Officer Charles was informed by Sgt. Lasiter that the search of the scene had been conducted by the laboratory technicians.

Sgt. Lasiter told Officer Charles the laboratory technicians had completed the search of the scene.

4. The juvenile suspect was found crouching behind some garbage cans in the alley by Officer Owens.

Officer Owens found the juvenile suspect crouching behind garbage cans in the alley.

5. It was determined by the officers that entrance was gained by the burglars through a side door.

The officers determined the burglars entered through the side door.

Index